Recent Titles in This Series

W0016930

The Evolution of
Haploid-Diploid Life Cycles

Lectures on Mathematics in the

LIFE SCIENCES

Volume 25

The Evolution of Haploid-Diploid Life Cycles

1993 Symposium on
Some Mathematical Questions in Biology
June 19–23, 1993
Snowbird, Utah

Mark Kirkpatrick
Editor

American Mathematical Society
Providence, Rhode Island

Proceedings of the 1993 Symposium on Some Mathematical Questions in Biology, held at the Annual Meeting of the Society for the Study of Evolution in Snowbird, Utah, June 19–23, 1993. This symposium was sponsored by the National Science Foundation under Grant DMS-9105171 02.

1991 *Mathematics Subject Classification.* Primary 92–06, 92D10, 92D15.

Library of Congress Cataloging-in-Publication Data

Symposium on Some Mathematical Questions in Biology (1993 : Snowbird, Utah)
    The evolution of haploid-diploid life cycles : 1993 Symposium on Some Mathematical Questions in Biology : June 19–23, 1993, Snowbird, Utah/Mark Kirkpatrick, editor.
    p.   cm. — (Lectures on mathematics in the life sciences, ISSN 0075-8523; v. 25)
    Includes bibliographical references.
    ISBN 0-8218-1176-2 (acid-free paper)
    1. Cell cycle—Congresses. 2. Cells—Evolution—Congresses. I. Kirkpatrick, Mark, 1956– .
II. American Mathematical Society.   III. Title.   IV. Series.
QH605.S957   1993   94-6316
574.87′623—dc20   CIP

Table of Contents

Introduction

The haploid-diploid alternation of generations was first described by Strasburger one hundred years ago, and modern population genetics was launched by Fisher seventy five years ago. Curiously, this most basic of genetic phenomena remains almost unexplored by this most obvious of approaches. A start was made on the remedy to this situation in a symposium convened at Snowbird, Utah in June 1993. The objectives were to review our current theoretical and empirical understanding and to stimulate a wider interest in the topic. This volume collects the contributions to that symposium.

While research on the evolution of genetic life cycles is still in infancy, interest in the topic has venerable roots. An anecdote about H.J. Muller shows concern with the evolution of diploidy long before this symposium. A dining companion watched as Muller ground a generous layer of black pepper onto his food. When the friend pointed out that pepper is mutagenic, Muller responded, "That's why we're diploid."

The view that diploidy is an adaptation to protect against deleterious recessive mutations is still popular. There seems to be more to the story, however, since not all organisms are diploid throughout their lives. Familiarity with the "higher" metazoans makes it easy to forget that other eukaryotes, including protists, algae, fungi, and mosses, show a diversity of life cycles. These include cycles in which diploidy is drastically reduced and others that feature a mixture of haploid and diploid phases. Explaining the existing diversity is one of the two big challenges that genetic life cycles present to evolutionary biology. The second is the origin of the alternation of generations. The solution to those riddles is likely to shed light on other fundamental problems that include the evolution of sexual reproduction and recombination, and the roles of mutation, segregation, and ecological selection in shaping the genetic system.

This volume is a synthesis of theoretical and empirical studies of the haploid-diploid life cycle. Graham Bell opens with a systematic review of the diversity of genetic life cycles in eukaryotes. He finds that diploid-dominated life cycles have arisen a surprising number of times, and suggests this trend has fundamental links to the evolutionary advantages of sexual reproduction. Alex Kondrashov's chapter tackles the important question of how the haploid-diploid alternation of generations may have originated. Using both genetic models and a survey of genetic systems in living eukaryotes, he develops two main points: that this complex arrangement could have evolved through a series of adaptive intermediates, and that genome-wide deleterious mutation could have provided the motive force. The following trio of chapters examine the subsequent evolution of a haploid-diploid life cycle using mathematical models, asking how various evolutionary forces determine the proportion of the life cycle spent in each phase. Jenkins and Kirkpatrick take up

Kondrashov's theme of deleterious mutation, and examine how the forces it generates interact with ecological selection acting on the haploid and diploid forms. Sarah Otto broadens the scope of evolutionary factors by considering effects of advantageous as well as deleterious mutations that appear throughout the genome. Her results make the important point that the masking effect of diploidy applies to good as well as bad mutations. Michod and Gayley's chapter considers life cycle evolution starting with the hypothesis that diploidy is critical to the repair of certain kinds of DNA damage. This perspective leads to predictions regarding evolution of many basic aspects of the breeding system (such as outcrossing and recombination) as well as the haploid-diploid cycle. In the final chapter, Véronique Perrot reviews the experimental data bearing on the assumptions of the theoretical models. She makes a persuasive case for future experimental work on algae and fungi that can be manipulated to shift the proportion of their life cycles that are spent in the haploid and diploid phases.

The new wave of interest in life cycle evolution reflected here has a polyphyletic origin. Two years ago, Alex Kondrashov and Jim Crow discovered that several research groups in North America and Europe had independently and simultaneously begun work on the topic. They organized an informal workshop at Madison in December 1991, and with tongue in cheek gave it the grand title of the "First International Congress on the Evolution of Haploid-Diploid Life Cycles." The success of that gathering inspired the symposium that convened at Snowbird a year later, and the participants thank Drs. Kondrashov and Crow for the impetus.

All of the participants are grateful to the symposium sponsors: the American Mathematical Society, the Society for the Study of Evolution, and the National Science Foundation. Their generosity allowed us to exchange recent results and, much more importantly, bring the topic to the attention of a wide audience. We thank Donna Salter, Donna Harmon, and Christine Thivierge at the AMS for their careful and good-humored work in organizing the symposium and producing this volume.

Mark Kirkpatrick
Department of Zoology
University of Texas
Austin, TX 78712 USA

Lectures on Mathematics in the Life Sciences
Volume **25**, 1994

The Comparative Biology of the Alternation of Generations

GRAHAM BELL

ABSTRACT. The life cycle of eukaryotes is a dual process in which growth and reproduction alternate with sexual fusion and meiosis. The vegetative processes may occur either in the haploid generation or in the diploid generation or in both. The problem of the alternation of generations is to explain the balance between growth in haploid and diploid individuals found in different groups of organisms. Although some organisms grow only as haploids or only as diploids, most multicellular organisms have both haploid and diploid growth, and the balance between the two varies at all phyletic levels. Haploid growth is primitive among eukaryotes, and diploid growth has evolved independently in several lineages. Diploid growth tends to be exaggerated in organisms with large size, complex organisation and specialized gametes, although there are many exceptions to the general rule. Two types of theory have been developed to account for the alternation of generations. The first type is based on attributes of vegetative ploidy. The best-known theory of this kind is that diploidy transiently reduces the expressed mutational load. The second type of theory is based on the relationship between the sexual and vegetative cycles. Among land plants, spores dispersing in air and gametes aggregating in water require different growth forms. More generally, there will be a tension between sexual selection for sexual performance and natural selection for vegetative performance. A simple genetic model shows that this theory seems to be capable of explaining the major comparative trends in the alternation of generations.

The life cycle of eukaryotes comprises a vegetative process of growth and reproduction, and a sexual process of fusion and restitution. The vegetative and sexual processes are qualitatively different, and it is essential to distinguish them clearly in order to understand the relationships among the life cycles of different groups of organisms. One of the main themes in the history of biology has been the elucidation of the nature of sexuality, and the separation of sexual from reproductive structures and events. Perhaps the single most important obstacle to

1991 Mathematics Subject Classification: 92D15

American Mathematical Association Symposium Joint Meeting of SSE, ASN, ASA Utah June 1993

Supported by a grant from the Natural Sciences and Engineering Research Council of Canada.

This paper is in final form and no version of it will be submitted for publication elsewhere.

progress has been that in many organisms, including ourselves, sex is obligately associated with reproduction, so that we tend to think of sex as a mode of reproduction. This has made it very difficult to understand the life cycles of the majority of organisms, in which sex and reproduction are more or less distinct. The situation was eventually clarified, at least in principle, by the development of modern concepts of genetics in the early years of the twentieth century. In practice, a good deal of confusion lingers on. Moreover, the correct interpretation of life cycles has raised fundamental theoretical questions which have not been solved, or even tackled very seriously, down to the present day. This Symposium marks a welcome change which brings life cycles once more into focus as one of the principal issues in evolutionary biology.

The Structure of Life Cycles

The life cycle is conventionally represented in a hundred elementary textbooks as a single series of events occurring during the lifetime of an individual, or of a set of different individuals related as parent and offspring, usually beginning and ending with a fertilized egg. I have developed a different system in which the life cycle as a whole is represented as a series of linked cycles, and specifically as two vegetative cycles of growth and reproduction linked to one another through a sexual cycle of meiosis and fusion. This is the system followed in Figures 1, 2 & 4, which form the backbone of this paper. The main purpose of this new representation is to distinguish more clearly between vegetative and sexual processes, because the problem of the alternation of generations is the problem of how and why the vegetative and sexual cycles are linked and related. I hope that although my diagrams will look unfamiliar at first sight they will soon enable the reader to understand the issues involved more clearly. The major features of the life cycles depicted in these figures are vegetative growth and reproduction; sexuality; heterogony; indirect development; the Steenstrup alternation of phases, or metagenesis; and the Hofmeister-Strasburger alternation of generations.

The Vegetative Cycle. The life cycle of eukaryotes begins with a unicellular, uninucleate spore. Many eukaryotes can proliferate through the detachment of relatively large masses of tissue by budding or fission or fragmentation, but in most cases it is also known that such organisms also produce unicellular spores, and the exceptions to the rule, if they exist, are few (Bell & Koufopanou 1991). The spore then grows in size, either by the enlargement of a single cell, or by the continued association of the products of cell division as a multicellular structure. The phenomena of growth and reproduction--and their counterpart, death--are universal features of the life cycle. In some organisms, such as euglenids and amoebas, they

FIGURE 1. The life cycle of eukaryotes. The life cycle is represented as linked sexual and vegetative cycles. (a) The dual life cycle, with direct development. This and the following three cycles are drawn as being haplontic; corresponding diplontic cycles could be drawn as easily. (b) Heterogony. The repetition of the vegetative cycle is indicated by a thick line. (c) Indirect development, with metamorphosis. (d) Steenstrup alternation of phases. (e) Hofmeister-Strasburger alternation of generations.

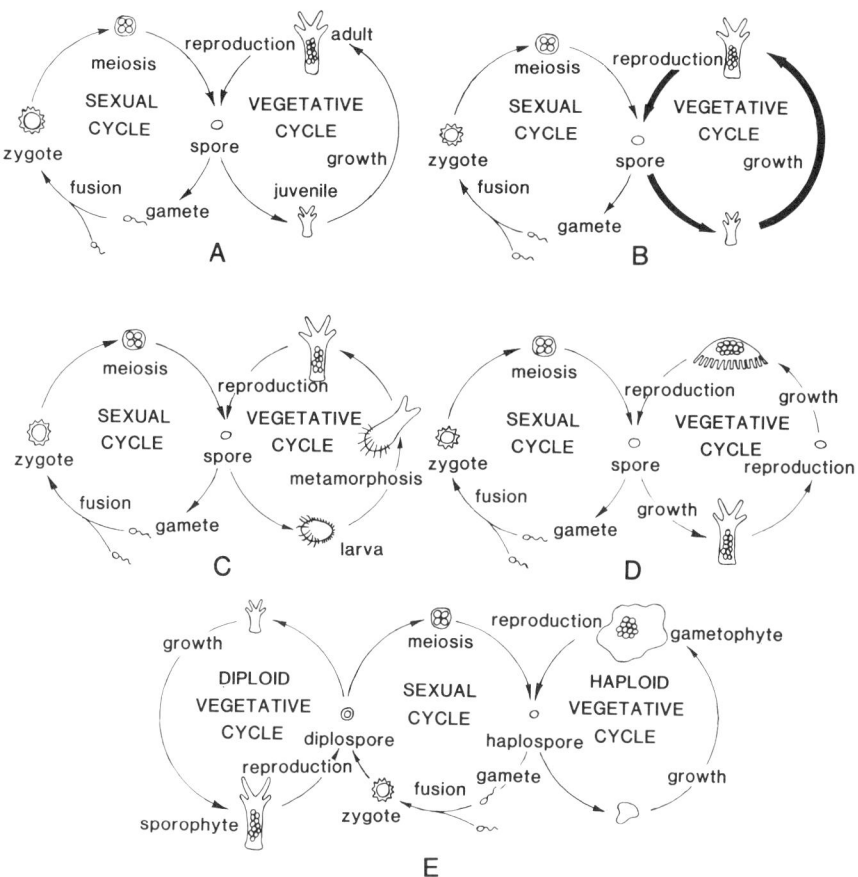

appear to be the only major features of the life cycle, which therefore comprises the vegetative cycle alone.

The Dual Life Cycle. In most eukaryotes the spore may be switched into either of two routes: it may enter the vegetative cycle, or it may enter the sexual cycle. These are quite different and incompatible fates: in the vegetative cycle, two cells (Figure 1a) are produced from one by growth, whereas in the sexual cycle one cell is produced from two by fusion. The life cycle of eukaryotes is therefore a dual process which is characterized by an alternation of some sort between vegetative and sexual development.

The sexual cycle is itself a dual process, since it always comprises two complementary events. Sexual fusion doubles ploidy as genomes from two independent lineages are brought together in the same nucleus. Meiosis is the sexual division, during which ploidy is halved, with new combinations of genes being produced by segregation and recombination. Fusion and meiosis are equally sexual phenomena; the sexual cycle necessarily involves an alternation between haploid and diploid nuclei.

Heterogony. Unlike growth and reproduction, sex is neither a necessary nor a universal feature of life cycles. Moreover, although the vegetative cycle may comprise the whole of the life cycle, it is clearly impossible that the sexual cycle should stand alone, without any phase of growth or reproduction, and purely sexual creatures are unknown. There is, rather, a continuum between organisms in which sex is obligately associated with reproduction, and in which a single cycle of growth therefore alternates with the sexual cycle, and those in which reiterated growth and reproduction are interrupted only rarely, if ever, by a sexual episode. Life cycles in which the vegetative cycle is usually iterated, and sex infrequent or sporadic, are said to be heterogonic (Figure 1b). All unicellular eukaryotes are heterogonic, if they are not entirely asexual; but heterogony also often occurs among large multicellular organisms.

Indirect Development. The simplest life cycles are simply a succession of unicellular spores. In many organisms, however, cell division gives rise to an aggregation of cells, which may be more or less differentiated to serve different functions, before this multicellular structure produces a new generation of spores. This implies a process of development from a smaller and simpler to a larger and more complex individual, the initial propagule and the mature reproductive adult being different to a greater or lesser extent. In many cases, development is punctuated by an abrupt change in appearance, for example from the ciliated, motile pelagic larva to the sessile benthic adult of animals such as echinoderms or sponges. This is only the extreme case of the phenotypic changes which are bound to occur during the development of any multicellular organism; nevertheless, where the change is very marked it is conventional, and useful, to distinguish it as a metamorphosis marking the transition from one phase to another of a vegetative cycle with indirect development (Figure 1c).

Steenstrup Alternation of Phases. The vegetative cycle may involve a regular succession of different types of individual developing from asexual propagules. The classical example is the proliferation of medusae from polyps in hydrozoans. It may be thought of as a process akin to indirect development in which metamorphosis is replaced by reproduction, with the development of individuals markedly different from their parents from spores or buds, rather than the production of a single transformed individual. There are many intermediate cases, such as the production of a single adult starfish from part of the body of a brachiolaria larva, where it is difficult to say whether growth, or reproduction by a single bud, is the more appropriate description. An alternation of morphological phases occurs regularly in several groups of metazoans, such as hydrozoans, trematodes and tunicates, in rhodophytes, and in scattered members of other phyla. It was first made widely known by Steenstrup (1845), and may therefore be called a Steenstrup alternation of phases. It is important to realize that the alternation of distinct reproductive types can occur within the vegetative cycle, and has no necessary connection with sexuality. The different types in such an alternation do not, therefore, differ in ploidy. Nevertheless, the two (or more) phases of a Steenstrup alternation are not equivalent with respect to the sexual cycle. The medusa of hydrozoans is the point of entry to the sexual cycle, producing haploid gametes by meiosis; the polyp is the point of exit from the sexual cycle, developing as a diploid creature from a zygote and producing diploid offspring. It seems appropriate to regard the medusa as a sexual individual, whereas the polyp is entirely asexual.

Hofmeister-Strasburger Alternation of Generations. The alternation of distinct types of individual seen in other organisms, such as ferns and seaweeds, is completely different in character. It occurs when two distinct vegetative cycles are separated by the sexual cycle; that is, when growth occurs both as a haploid individual, after meiosis and before fusion, and as a diploid individual, after fusion and before meiosis. This is, indeed, the general case, and haplontic cycles (growth of haploid individuals only) and diplontic cycles (growth of diploid individuals only) mark the two extremes of a continuum, with many organisms having intermediate life cycles. I shall call this the Hofmeister-Strasburger alternation of generations, after the two nineteenth-century biologists who were primarily responsible for the modern interpretation (Hofmeister 1851; Strasburger 1894).

Because Steenstrup worked on marine invertebrates and Hofmeister on conifers, the term "alternation of generations" is liable to mean different things to zoologists and botanists, which is why I have distinguished the two different concepts by the names of their most prominent early exponents, and refer to the Steenstrup alternation as involving phases rather than generations. It is possible to have both types of alternation within the same life cycle; this is often the case in rhodophytes, for example, whose life cycle is discussed in the next section.

Growth as a haploid individual after meiosis results in a gamete-producing adult, to identify which there is no alternative to the botanical term gametophyte. Growth as a diploid individual after fusion results in a mature spore-producing adult, the sporophyte. The distinction that I have made here between gametes and spores is conventional. It seems equally acceptable, and in my view clearer and more consistent, to recognize that both gametophytes and sporophytes produce spores, defined as uninucleate propagules having the same ploidy as their parent. Gametophytes produce haplospores, which either differentiate into a new haploid individual, or enter the sexual cycle and undergo fusion to create a diploid individual. Sporophytes produce diplospores, which either differentiate into a new diploid individual, or enter the sexual cycle and undergo meiosis to create a haploid individual. The gametophyte is often referred to as the sexual individual, and the sporophyte as the asexual individual. I think that this usage is subjective and misleading. Rather, both gametophyte and sporophyte should be regarded as vegetative phases which stand in different relationships to the sexual cycle. This makes it clear that the interpretation of the gametophyte-sporophyte alternation hinges on the relationship between the vegetative and sexual cycles. This is the issue which is the subject of our symposium. Why do some organisms grow as haploids, some as diploids, and others as both?

Comparative Distribution of Life Cycles

The purpose of this section is to provide a very brief overview of the Hofmeister-Strasburger alternation of generations throughout the eukaryotes, and to comment on some of the issues which are involved in using this information in comparative tests of theoretical interpretations of life cycles. The primary literature on life cycles is so vast that it is impossible to review it adequately here; I shall instead deal in broad generalizations, ignoring a million interesting details and exceptions. Indeed, no detailed review of life cycles has been attempted. The best point of entry to the literature is Margulis et al (1990); Bold & Wynne (1978) and Lobban & Wynne (1981) are useful sources for the autotrophic groups. I have previously given a brief comparative review of the subject (Bell 1982). Figure 2 is a pocket summary of eukaryote life cycles.

There are no isolated sexual cycles. Life cycles in which sex occurs without growth do not occur. Isolated vegetative cycles, on the other hand, are commonplace. They usually arise secondarily by the loss of sexuality; but the lack of sex in groups such as Euglenida may be primitive.

Most multicellular organisms have both haploid and diploid growth. Growth during the vegetative cycle can be achieved in either of two ways.

First, it may involve the continued growth of a single individual. This is the only sort of growth represented in the Figures. Among multicellular forms, there is no growth in the diploid cycle among charophytes, conjugaphytes, zygomycete and

FIGURE 2. Life cycles among eukaryotes. Life cycles are represented by simplified versions of Figure 1(e), from which all illustration of structures, other than haplospore (single circle) and diplospore (concentric circles), has been omitted. Individual growth is expressed crudely by four categories: no cycle, no growth; small cycle, unicells, filamentous or small crustose thalli, coenobia or coenocytes; medium cycle, parenchymatous thalli with extensive division of labour; large cycle, large complex highly-integrated individuals. These four levels in both haploid and diploid vegetative cycles give rise to 16 possible life cycles, each with a different balance between haploid and diploid growth. Abbreviations are as follows: H, haploid vegetative cycle; h, haplospore; S, sexual cycle; f, gamete fusion; m, meiosis; D, diploid vegetative cycle; d, diplospore. Arrows indicate direction of development.

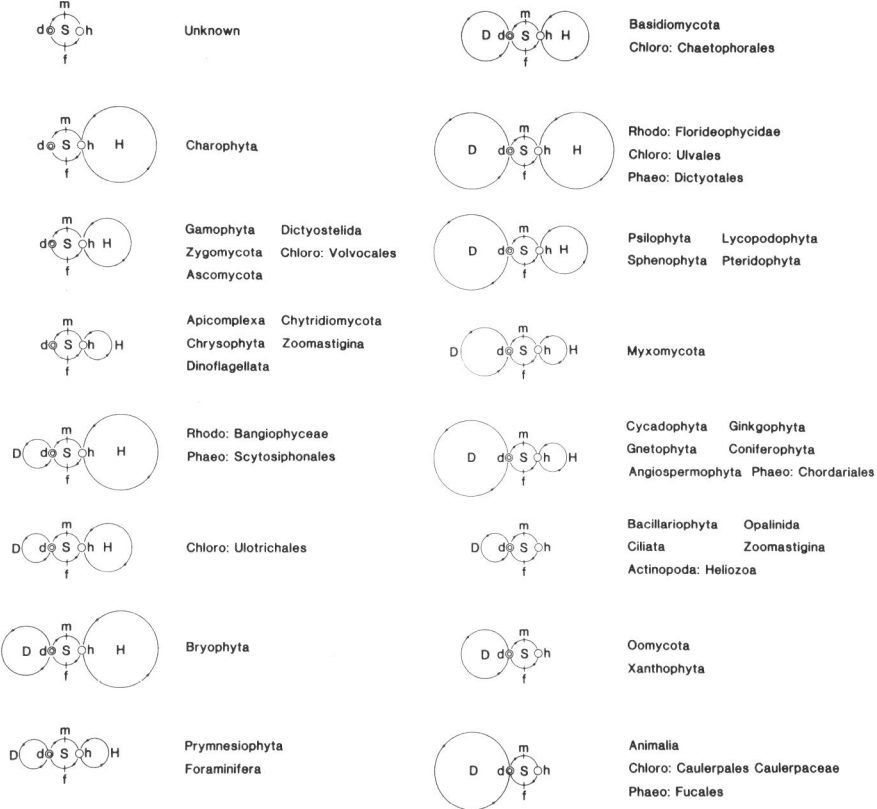

Unknown

Charophyta

Gamophyta   Dictyostelida
Zygomycota   Chloro: Volvocales
Ascomycota

Apicomplexa   Chytridiomycota
Chrysophyta   Zoomastigina
Dinoflagellata

Rhodo: Bangiophyceae
Phaeo: Scytosiphonales

Chloro: Ulotrichales

Bryophyta

Prymnesiophyta
Foraminifera

Basidiomycota
Chloro: Chaetophorales

Rhodo: Florideophycidae
Chloro: Ulvales
Phaeo: Dictyotales

Psilophyta   Lycopodophyta
Sphenophyta   Pteridophyta

Myxomycota

Cycadophyta   Ginkgophyta
Gnetophyta   Coniferophyta
Angiospermophyta Phaeo: Chordariales

Bacillariophyta   Opalinida
Ciliata   Zoomastigina
Actinopoda: Heliozoa

Oomycota
Xanthophyta

Animalia
Chloro: Caulerpales Caulerpaceae
Phaeo: Fucales

ascomycete fungi, dictyostelid slime moulds and coenobial Volvocales. There is no growth in the haploid cycle among animals and some water-moulds and seaweeds. In all other cases, both the haploid and diploid vegetative cycles involve growth and reproduction. Nevertheless, isomorphic cycles in which haploid and diploid individuals are morphologically indistinguishable, or nearly so, occur only in basidiomycete fungi, many florideophycean rhodophytes and foraminiferans, and some phaeophytes and chlorophytes. They also occur in the unicellular foraminiferans. In other taxa there is a tendency, often very pronounced, for the haploid and diploid individuals to differ markedly in size and structure.

Secondly, growth may involve reproduction through the heterogonic repetition of either or both vegetative cycles. This is invariably the case among unicellular forms, but is also widespread among multicellular organisms, where either the haploid or the diploid cycle, or both, may be iterated asexually. Moreover, when either cycle includes a Steenstrup alternation, it is often the case that a given morphological phase can recycle itself. The life cycles of many organisms are thus much more complicated than they are represented as being in simple stylized diagrams. In such cases, it is seldom if ever known how often lineages switch from vegetative to sexual expression. The total amount of growth achieved in either vegetative cycle is then unknown, and the balance between haploid and diploid growth may not be fairly represented by the relative size of haploid and diploid individuals.

The balance of haploid and diploid growth varies at all phyletic levels. Extreme life cycles tend to be rather uniform; when there is a great disparity in growth between haploid and diploid individuals, this disparity characterizes all members of a phylum. Thus, all charophytes and zygomycetes are strictly haplontic; all animals and ciliates are strictly diplontic. It may be difficult to evolve a new vegetative cycle, once the old one has entirely or almost entirely disappeared. In phyla where both haploid and diploid individuals can grow substantially, life cycles may vary widely. In at least three phyla--Chlorophyta, Phaeophyta and Rhodophyta--there are extensive series of forms intermediate between completely, or nearly completely, haplontic taxa at one extreme and diplontic taxa at the other. Even at much lower taxonomic levels there may be a great deal of diversity. An example is provided by the monostromatic (thallus a single layer of cells) chlorophytes formerly assigned to the single genus Monostroma (see Tanner 1981, Table 6.1). All are morphologically similar and occupy similar habitats in fresh and brackish water. Monostroma itself has a heteromorphic alternation between a large multicellular leafy gametophyte, the familiar seaweed, and a small saccate sporophyte, the so-called "Codiolum" phase, originally described as a separate genus. Ulvaria has an isomorphic alternation, with both gametophyte and sporophyte developing as large leafy thalli. In Kornmannia the sporophyte is large and leafy, while the gametophyte is a minute prostrate disc. Other forms seem to be entirely asexual, and either grow only as leafy thalli, or show a Steenstrup alternation between leafy and Codiolum phases; the retention of the morphological alternation when sex has been lost is particularly striking. Many other instances are known in which related forms have struck a different balance between haploid and diploid growth, or in which the loss of the sexual cycle restricts growth to one or the other generation, with or without the retention of an alternation of morphological phases. Finally, many eukaryotic microbes (including those most often used in the laboratory) seem to vary indefinitely between haplonty and diplonty. Unicellular yeasts may reproduce indefinitely either as haploid or as diploid cells, depending on whether they are effectively homothallic. Chlamydomonas normally reproduces as a haploid, but zygotes occasionally germinate as diploid cells, having failed to undergo meiosis, and will continue to proliferate as diploids. It is even possible to construct stable diploids by artificially fusing protoplasts.

The uniformity of life cycles in groups such as animals suggests that when either the haploid or the diploid vegetative cycle is lost selection may be unable to restore it because of a lack of relevant variation. In taxa with less extreme life cycles, however, the balance between haploid and diploid growth often varies greatly among related taxa. In these taxa, therefore, it is likely that the life cycles we observe are actively maintained by short-term selection. This implies that selection may favour any balance between haploid and diploid growth, depending on circumstances; it is the object of theory to identify what these circumstances are.

Haploidy is primitive among eukaryotes. The evolution of life cycles cannot yet be traced in detail, because there is little agreement regarding some of the earliest events in eukaryote history. Three reasonable, but conflicting, points of view are given in Figure 3. However, some conclusions seem to be entailed by almost any reasonable phylogeny.

The most economical interpretation of the cladograms is that haplonty is primitive, with growth in the diploid generation evolving subsequently and independently in several lineages. The earliest-diverging freeliving groups in phylogenies estimated through the analysis of rRNA sequences are almost invariably euglenids, amoebomastigotes and the cellular and plasmodial slime moulds. The slime moulds have well-documented sexual cycles; their vegetative cells are haploid amoebas. Diploid growth was subsequently evolved in several clades. All three phylogenies agree on four of these lineages: animals, "greens" (the clade which includes chlorophytes, gamophytes and plants), basidiomycetes and ciliates. Two of the phylogenies add rhodophytes (which the third did not sample); one also adds a Bacillariophyta-Oomycota clade, and the other the Prymnesiophyta. No doubt

FIGURE 3. The phylogeny of life cycles. Extant taxa are labelled according to whether growth is haploid (H), diploid (D) or both (HD). One possible parsimonious assignment of character states to nodes is also indicated. Three phylogenies based on sequence analysis of 16S-like rRNAs are shown, all of which have been proposed in the recent literature. Model I, with early appearance of animals and rhodophytes, is consistent with Hendriks et al. (1991), Douglas et al. (1991) and Martin, Somerville & Loiseaux-de Goer (1992). Model II, with late appearance of animals and rhodophytes, animals being a sister-group to green plants, is consistent with Elwood, Olsen & Sogin (1986), Sogin & Elwood (1986), Qu et al. (1988), Baroin et al. (1988), Perasso et al. (1989), and Lenaers et al. (1989, 1991). Model III, with animals and fungi as sister-groups, is consistent with Wainwright et al. (1993). Some well-supported features of these and other phylogenies have been incorporated into the three models, though in some cases the taxa involved were not sampled by the authors listed above. These include the basal position of euglenids; ciliates and dinoflagellates as sister-groups; and the branching-order of "greens" (see Devereux, Loeblich & Fox, 1990).

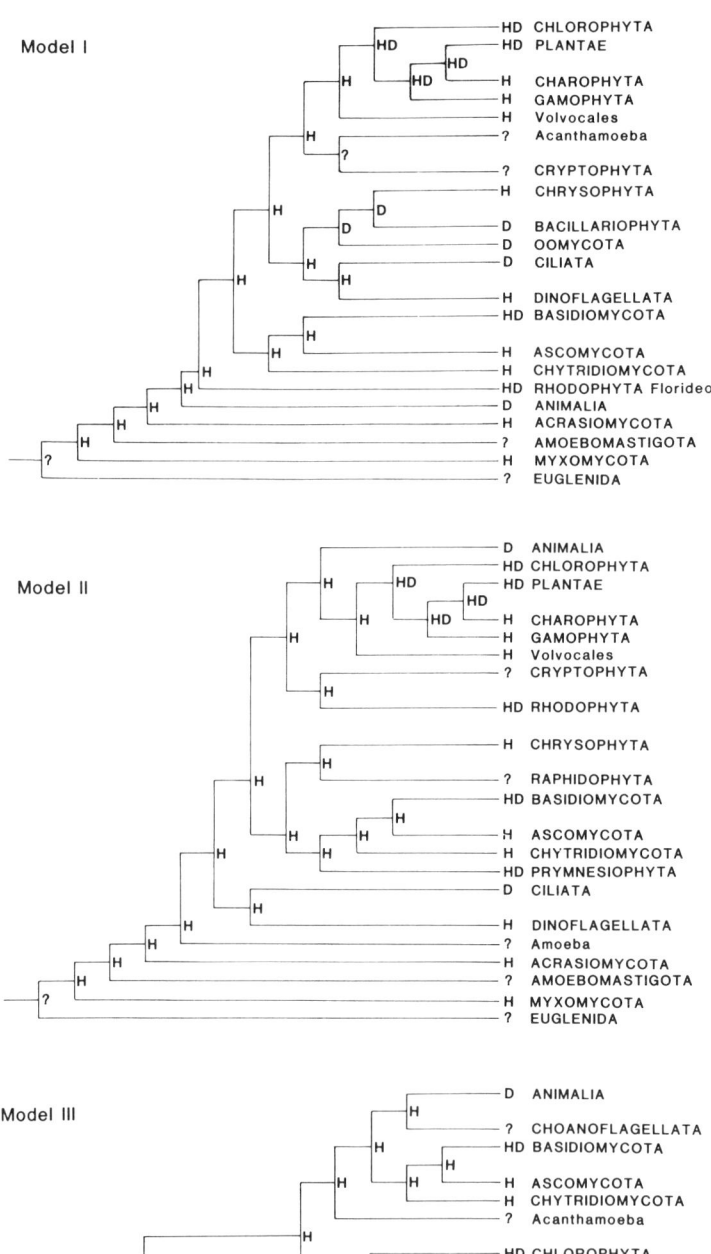

phylogenetic schemes will change again in the near future; it seems very likely, however, that whatever scheme is eventually adopted will show several lineages independently evolving diploid growth.

Whether diplontic lineages have ever reverted to haplonty is less clear. If the branching order of the clade of greens is Volvocales-Chlorophyta-Gamophyta-Charophyta-Plantae, then the gamophytes might be an example. It is economical to infer that they arose from an ancestor with diploid growth, since if this ancestor were haplontic we must postulate that diploid growth evolved twice, in Chlorophyta and in Charophyta-Plantae, rather than only once, in the ancestor of the Chlorophyta-Gamophyta-Charophyta-Plantae clade. One of the phylogenetic schemes also suggests that the haplontic Chrysophyta are derived from diploid ancestors, but this is not supported by the other phylogeny which samples chrysophytes. Clearly, this issue can be resolved only when a basic eukaryote phylogeny has been established.

The question of whether diplontic life cycles can ever revert to haplonty has one curious echo in animals. All animals are diplonts, yet there are many species in the phyla Arthropoda and Rotifera where the males are haploid, developing from unfertilized eggs. This curious phenomenon of arrhenotoky shows that there is no necessary bar to haploid development, even in taxa which lack a regular haploid vegetative cycle.

There are correlated shifts in life cycle, somatic organization and gamete specialization. There is a general tendency for diploid growth to be associated with a large and complex soma and with highly specialized gametes. The groups in which growth of the individual is achieved largely or entirely in the diploid vegetative cycle include most of the largest and most complex organisms: animals, land plants and large seaweeds. Many seaweeds also grow as haploid individuals, but otherwise the largest and most complex organisms with haploid growth are charophytes and bryophytes. In most diplonts, the gametes are highly specialized cells which are quite different from vegetative cells and which are incapable of growth; gamete fusion involves a minute motile sperm and a massive immotile egg and is said to be oogametic. In many haplonts, fusion occurs between little-modified vegetative cells, both gametes being motile whether they are similar (isogamety) or dissimilar (anisogamety) in size; moreover, reproduction often involves zoospores which resemble gametes but are unable to grow. There are, however, many exceptions to these trends, which can be described in more detail for the grades and clades within which life cycles vary.

Eukaryotic microbes. Unicellular eukaryotes may be either haplontic or diplontic, or alternate between haploid and diploid vegetative cycles. Freeliving forms in which the gametes closely resemble unmodified vegetative cells (chlorophytes, chrysophytes, dinoflagellates and dictyostelids) are almost invariably haplontic, and are generally isogametic. The only freeliving unicellular diplonts which produce flagellated gametes are the centric diatoms; these, however, are oogametic, and diatoms do not produce zoospores. In the diplontic ciliates and pennate diatoms, the entities which fuse are diploid gamonts (gamete-bearing individuals) resembling vegetative cells, haploid protoplasts or nuclei being formed subsequently by meiosis and fusing within the structure formed by the paired gamonts. Filamentous forms with gametangial copulation and immotile or amoeboid gametes may be haplontic (gamophytes, zygomycetes and ascomycetes) or diplontic (oomycetes). Chytrids are haplonts in which the vegetative thallus is a large multinucleate structure quite different from the gametes; they also produce zoospores, however, which are morphologically very similar to gametes. In dictyostelid slime moulds the gametes are amoeboid, but closely resemble the haploid trophic vegetative amoebas.

The most familiar freeliving unicellular eukaryotes with a regular alternation of generations are foraminiferans, where the haploid and diploid generations are nearly isomorphic. Most foraminiferans have gamontic copulation, but some produce specialized flagellated sperm. At least some prymnesiophytes alternate between a motile pelagic generation which is diploid and an immotile benthic cell or minute filament which is haploid; they produce zoospores and flagellated freeswimming gametes.

Among unicellular eukaryotes generally, therefore, forms which produce flagellated gametes that are unspecialized in that they resemble vegetative cells invariably have haploid growth, and most are strictly haplontic. The production of specialized gametes is associated with diploid growth in foraminiferans and centric diatoms. No such generalization applies to groups in which the fusing entities are gamonts or gametangia, which may be haplontic or diplontic or have an alternation of generations. In such forms, the movement and fusion of amoeboid gametes (as in gamophytes or foraminiferans) or protoplasts (as in zygomycetes) or nuclei (as in ciliates) occurs within the space formed by the fused gamonts or filament cells. The generalization that forms with unspecialized gametes are haplontic breaks down, therefore, when there is a severe spatial restraint on gamete fusion.

Phaeophyta. All phaeophytes except some species of Fucus have both haploid and diploid growth (Figure 4a). The simplest alternate between haploid and diploid filamentous phases, which may be isogametic (Ectocarpales) or markedly anisogametic or oogametic (Tilopteridales, Sphacelariales). The Dictyotales have an isomorphic alternation of large thalli; they are oogametic. The haploid cycle is most prominent in the Scytosiphonales, where the gametophyte is a large cylindrical or flattened thallus, whereas the sporophyte is crustose or filamentous; they are nearly isogametic. Most large and somatically complex forms (e.g. Desmarestiales, Laminariales) have large diploid thalli, the gametophyte being reduced to a minute filament. They are mostly oogametic, though some (e.g. Chordariales) are isogametic. The Fucales are the extreme expression of this trend, the only haploid cells being gametes, or at most the cells produced by two or three mitoses following meiosis; they are oogametic. The phaeophytes therefore exemplify the general tendency for diploid growth, somatic size and complexity, and gamete specialization to vary together.

Chlorophyta. The chlorophytes are also extremely variable (Figure 4b). Unicellular and filamentous chlorophytes are mostly haplontic, or nearly so, and are isogametic or somewhat anisogametic. Two notable exceptions are Oedigoniales and the coenobial Volvocales, which are oogametic haplonts. Some Oedigoniales produce freeswimming male gametes resembling zoospores, but in others the male is a dwarf filament which attaches itself adjacent to an oogonium in a large female filament. Coenobial Volvocales produce packets of male gametes which are essentially dwarf colonies, and which penetrate the wall of the female colony before liberating the individual sperm inside. In both cases, therefore, there is again the suggestion that the production of specialized gametes by haplonts is associated with a spatial restriction on gamete fusion. There are many chlorophytes which have an isomorphic alternation of relatively large and complex thalli; most are anisogametic, though some (e.g. Chaetophorales) are isogametic, and others (e.g. Prasiolales) are oogametic.   Only certain Caulerpales seem to be diplontic; they have large rhizomatous foliose thalli, and are anisogametic.

"Greens". Chlorophytes lie at the base of a clade of chlorophytes-gamophytes-charophytes-plants. This clade supplies the classical instance of a tendency for diploid growth to be associated with large size and complex organization. The charophytes are haplontic, the zygote being the only diploid structure in the life cycle. The leafy plant body of bryophytes is haploid, the sporophyte being a fairly large structure borne on the gametophyte.   Many vascular plants, such as pteridophytes, have a large leafy sporophyte and a much smaller filamentous or thalloid gametophyte. In conifers and seed plants there is no freeliving haploid generation. The sequence is in this case almost certainly a temporal one, with charophytes appearing first in the geological record, bryophytes next, and the vascular land flora last. There is therefore a very clear tendency for the diploid vegetative cycle to replace the haploid cycle as the principal route of growth with the evolution of larger size and more highly derived means of gamete transfer.

Rhodophytes. Rhodophytes have extraordinary life cycles (Figure 4c). The classical pattern, found in most orders of Floridophycidae, involves an alternation between two diploid phases and a haploid gametophyte. The nonflagellated male gamete fertilizes a sessile oogonium, which develops into a mass of diploid tissue, the carposporophyte, sessile on the gametophyte.   This produces diploid carpospores which on germination develop into a freeliving tetrasporophyte isomorphic with the gametophyte. There is therefore both a Steenstrup alternation of phases (carposporophyte and tetrasporophyte within the diploid vegetative cycle) and a Hofmeister-Strasburger alternation of generations (carposporophyte-tetrasporophyte with gametophyte) within the life cycle. In many cases, however, the haploid thallus is much larger than any diploid structure. This may involve the reduction of the diploid cycle to the carposporophyte alone, or the reduction of the tetrasporophyte to a prostrate filamentous or crustose individual.   In Bangiophycidae there is only a single diploid phase, which is a minute filament. The only forms in which the diploid individual is the larger are some Acrochaetiales, in which both the haploid and the freeliving diploid plants are small and inconspicuous filaments. There seems little indication in rhodophytes of any tendency for larger and more complex forms to emphasize diploid growth, and little variation in the degree of specialization of the gametes.

FIGURE 4. Variation of life cycles within phyla. These three diagrams illustrate the range of life cycles found in: (a) Phaeophyta; (b) Chlorophyta; (c) Rhodophyta. Vegetative form is illustrated by pictograms as follows:

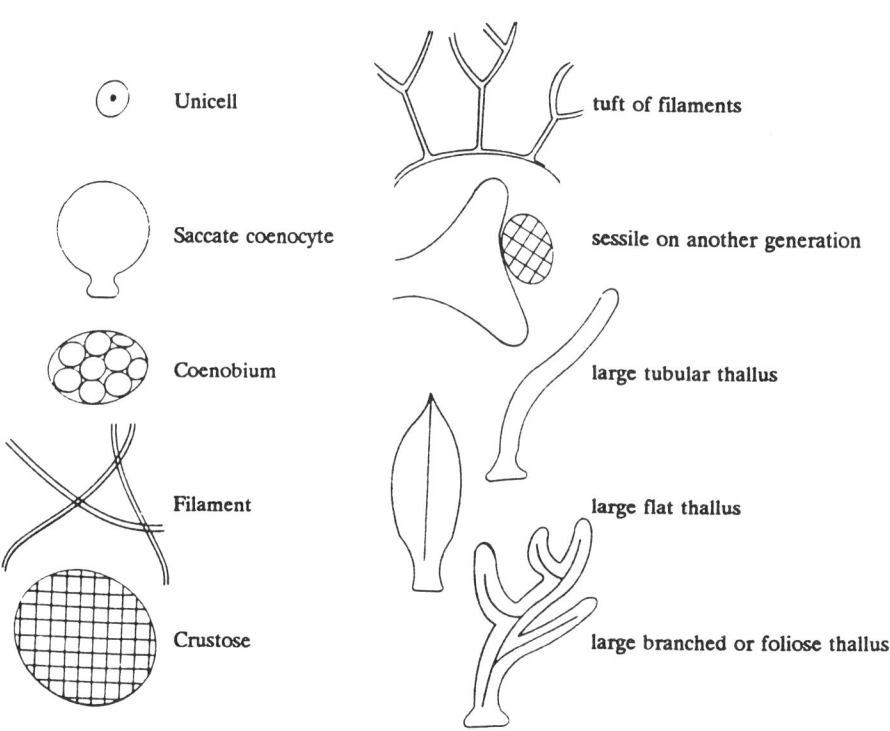

Unicell

tuft of filaments

Saccate coenocyte

sessile on another generation

Coenobium

large tubular thallus

Filament

large flat thallus

Crustose

large branched or foliose thallus

Gametic specialization is illustrated as follows:

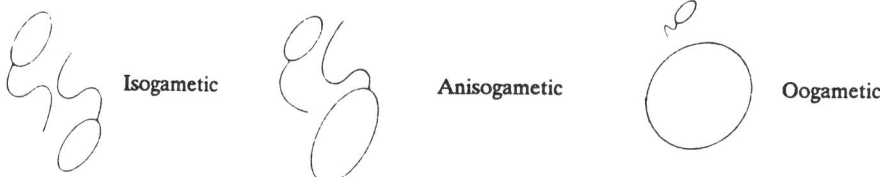

Isogametic

Anisogametic

Oogametic

Note the occurrence of two diploid phases in the life cycles of many rhodophytes. The gametic and vegetative stereotypes which are shown here are intended to be typical, but often conceal a great deal of variation within taxa. In particular, gamete dimorphism often varies among and within orders of chlorophytes and phaeophytes which have similar life cycles.

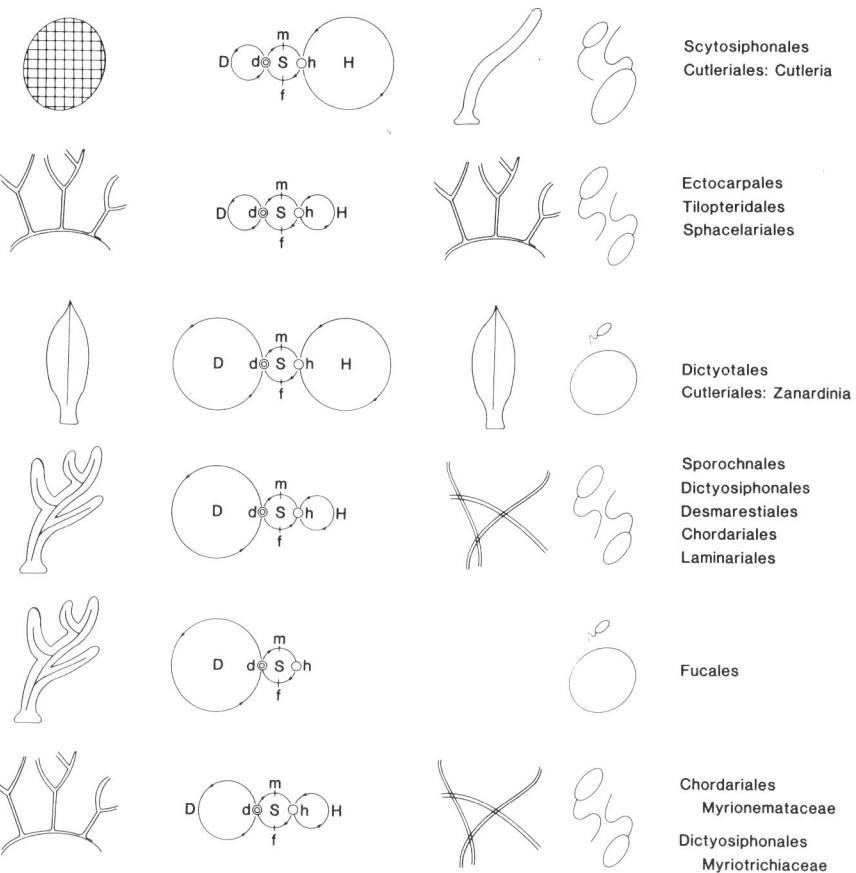

Scytosiphonales
Cutleriales: Cutleria

Ectocarpales
Tilopteridales
Sphacelariales

Dictyotales
Cutleriales: Zanardinia

Sporochnales
Dictyosiphonales
Desmarestiales
Chordariales
Laminariales

Fucales

Chordariales
  Myrionemataceae

Dictyosiphonales
  Myriotrichiaceae

Figure 4(a): Phaeophyta

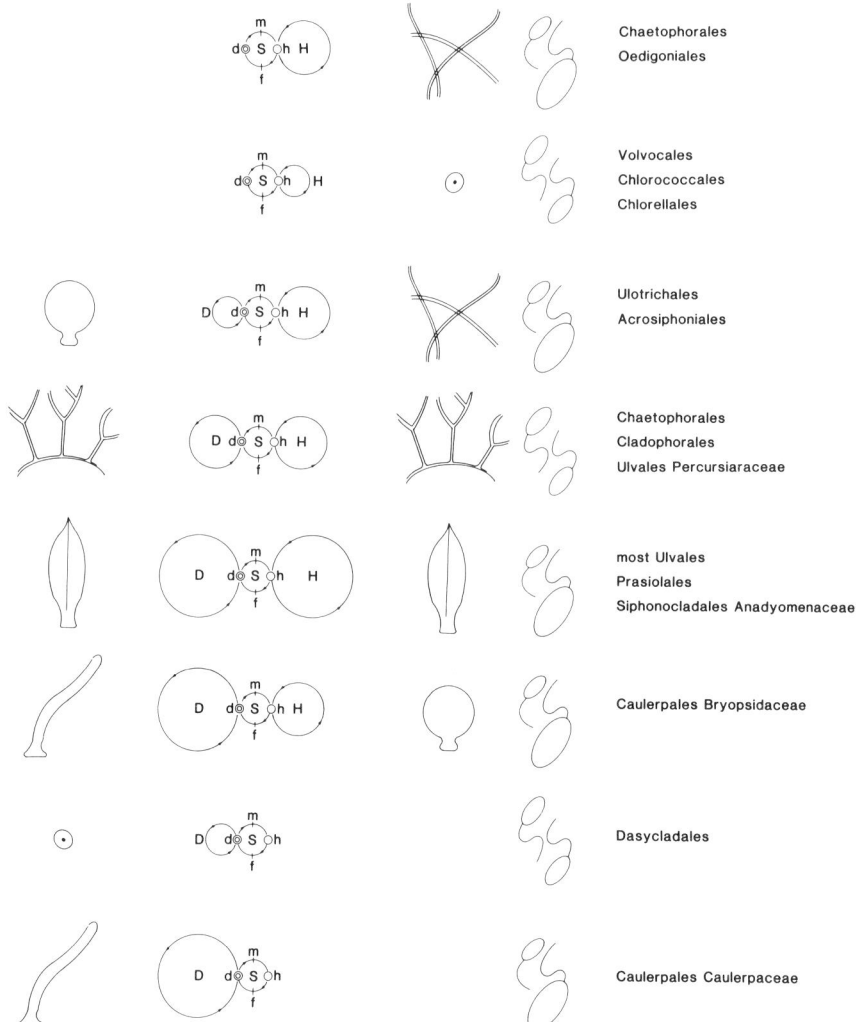

Figure 4(b): Chlorophyta

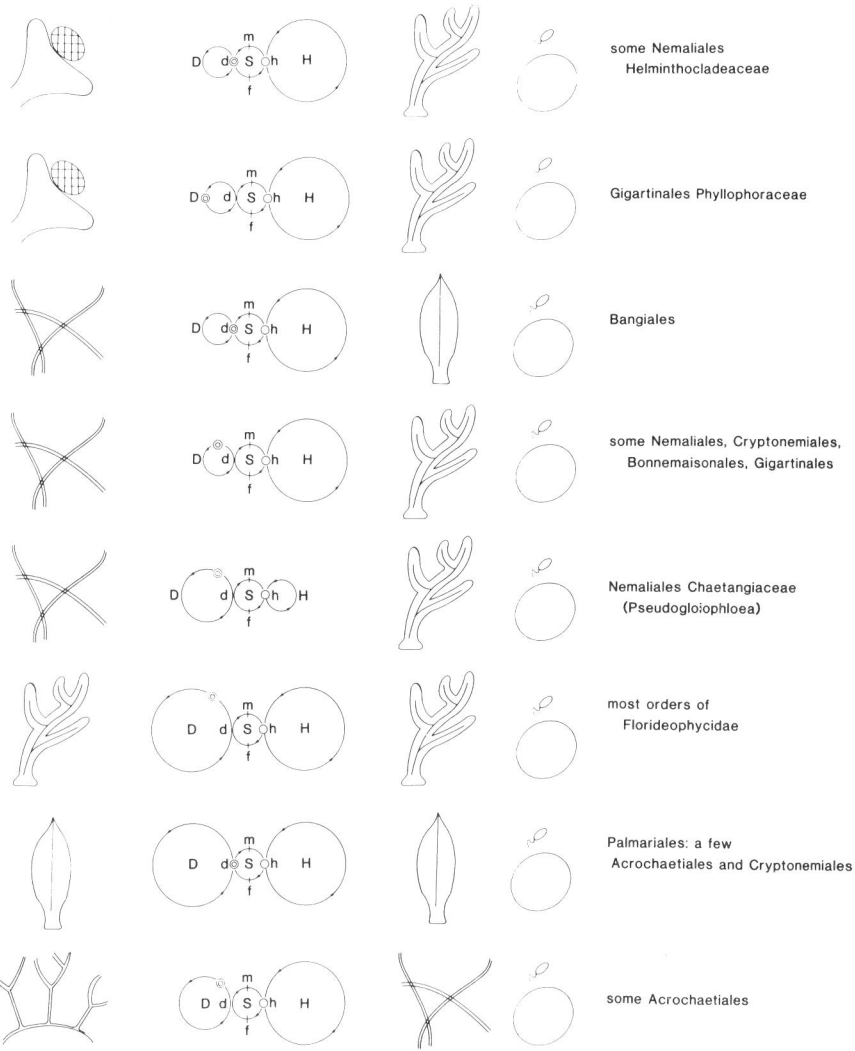

Figure 4(c): Rhodophyta

Theories of the Life Cycle

The elucidation of life cycles in the later part of the nineteenth century was quickly overtaken by developments in physiology, genetics and experimental embryology, and is nowadays relegated to a minor place in biology curricula. A great deal of knowledge has accumulated during the twentieth century regarding the life cycles of particular organisms, but evolutionary interpretations of this enormous mass of material have been largely concerned with continuing the phylogenetic speculations of a hundred years ago, rather than developing functional and adaptive theories. Knowing the phylogeny of life cycles is essential if their evolution is to be interpreted correctly. However, a functional approach is equally essential. In some groups, evolution may be so narrowly constrained by a lack of relevant variation that the life cycle is non-optimal for most members of the group, having originally been fixed in different circumstances long ago. But in many groups, and especially in groups where both haploid and diploid individuals grow, there is a great deal of variation in life cycles, even within taxa of low taxonomic rank. In these groups it is reasonable to suppose that selection continues to act on the balance between haploid and diploid growth, and that current life cycles represent nearly optimal states which require functional explanation. A sound theoretical basis for understanding the adaptedness of life cycles was not developed when they were one of the main objects of scientific enquiry, and is still lacking today (but see Valen et al. 1992.

Two quite different types of theories of the life cycle can be distinguished. The first comprises theories which deal exclusively with ploidy; in most cases, such theories attempt to explain why most large and familiar organisms are diplontic or nearly so. Theories of this sort do not require an alternation of generations, since they can be applied with equal force to exclusively asexual life cycles. The second category comprises theories which are based on the relationship between the sexual and the vegetative cycles. Such theories regard the alternation of generations as being a problem in its own right, which requires an explanation separate from the consequences of different ploidies among asexual organisms.

Theories based on attributes of vegetative ploidy. Increasing the number of genomes per nucleus, for whatever reason, may have divers physiological, developmental and genetic consequences, all of which have been advanced as theories of the life cycle.

Ecophysiological theories. The relationship between genome mass and cell mass is well known (see Cavalier Smith 1985). Larger cells have more copies of the genome in order to maintain levels of transcription adequate to support their greater total metabolic rate (Bell 1989). Large unicellular eukaryotes such as euglenids and amoebas are therefore polyploid or multinucleate, or both, or have evolved some equivalent but more sophisticated system, such as the nuclear dimorphism of ciliates and foraminiferans. It is easy to understand why large cells require more copies of the genome, but what is to prevent small cells from being polyploid? Lewis (1985) has pointed out that in such organisms the genome represents a large, noncirculating fraction of total cell phosphorus, and is therefore a substantial investment for a small autotrophic unicell living in oligotrophic conditions. Haploidy will be favoured by selection because it conserves nutrients, allowing growth at low nutrient concentrations; diploidy, or higher ploidy, permits higher rates of growth through the wasteful use of nutrients when nutrients are abundant. This argument applies only to unicells, because multicellular organisms can regulate the allocation of resources to DNA by adjusting cell size without affecting body size. With this limitation, admittedly a rather severe one, this is an admirably simple theory which might explain, for example, why large unicellular heterotrophs are generally diploid (e.g. ciliates) or have diploid growth (e.g. foraminiferans), while most small photoautotrophic algae are haplontic. It is also easy to test experimentally, by estimating the growth parameters of isogenic haploid and diploid strains of microbes such as yeast or <u>Chlamydomonas</u> over a range of nutrient concentrations.

Developmental theories. There is a persistent undertone in theories of life cycles to the effect that diploidy predominates among large organisms because it is for some reason essential for complex somatic differentiation (e.g. Maynard Smith 1978, p. 9). The difficulty of specifying what the reason might be presumably reflects a weakness in the theory. Haplontic organisms such as charophytes can be rather large and complex. The haploid vegetative generations of many phyla of multicellular organisms are as large and complex, or larger and more complex, than the diploid phases. The haploid males of hymenopterans and mites are comparable in structure with the diploid females. This hypothesis should be abandoned.

Genetic theories. The sexual basis of the alternation of generations has always attracted the attention of population geneticists. Paradoxically, the theories that they have developed do not involve the sexual cycle directly. Instead, they are based on the properties of asexual heterozygotes.

By far the most popular theory of this sort has been that deleterious mutations are complemented in diploids. Although the idea is quite old (Muller 1932; Crow & Kimura 1965) it has recently been developed formally at some length (Kondrashov & Crow 1991; Perrot, Richerd & Valero 1991); because it will be discussed extensively in this symposium I shall not describe it in detail here. The fact that individuals of any grade of structure may be haploid is not immediately fatal to the theory. The evolution of a diploid vegetative cycle in taxa where none exists may be severely constrained by an absence of relevant variation, although the fact that so many taxa exist does seem difficult to explain. A more subtle point is that that when diploidy does evolve the process should be irreversible. A diploid vegetative cycle will cause a transient reduction in the expressed genetic load, but thereafter deleterious recessive mutations will accumulate, precisely because they are complemented by unmutated alleles. A reversion to haplonty would result in the immediate expression of these deleterious genes, just as they are expressed by inbreeding. In fact, although well-supported phylogenies suggest that diploid growth has evolved from haplonty on several occasions, there are few examples of the reverse transition, and none that are beyond dispute. Arrhenotoky is more troublesome, since the haploid males should on this theory be mostly inviable. It may be argued that where a cycle which includes a regular haploid phase has evolved, the genetic load will have been reduced; but this seems to threaten the basis of the theory. More fundamentally, it is not clear how the theory can begin to address the great diversity of life cycles which represent a balance between haploid and diploid growth.

Alternatively, it has been suggested that diploid populations may evolve more rapidly than competing haploids. Lewis & Wolpert (1979) have argued that the acquisition of new functions by haploids will usually involve gene duplication followed by sequence divergence, whereas in diploids the divergence of the two copies of a gene can occur immediately, to be followed by duplication. It can be shown that the latter process permits a more rapid acquisition of novel functions. Whether or not this is the general case may be doubted. Gene duplication in haploids may generate an immediate advantage by increasing the rate of production of an inefficient enzyme, one copy of the gene being subsequently modified so as to become more efficient in utilizing the new substrate. Several examples of such a process have been documented in bacteria (see Mortlock 1984). Nevertheless, there is some experimental evidence that diploid lineages evolve faster than haploids in laboratory cultures of yeast (Paquin & Adams 1983). The main weakness of the theory is, again, that by attempting to identify an unequivocal advantage of diploidy it makes it difficult to interpret comparative patterns.

Theories based on the relationship between the sexual and vegetative cycles. The second category of theories is quite different. Neglecting the physiological consequences of differences in ploidy, it attempts to explain why growth should occur between meiosis and fusion to produce a haplontic cycle with a zygotic meiosis; or between fusion and meiosis to produce a diplontic cycle with a gametic meiosis; or in both generations, to produce an alternation of generations with a "sporic" meiosis. It therefore hinges on interpreting the relationship between the sexual and vegetative cycles.

Gamete and spore dispersal. The oldest theory of this kind was put forward by Bower in 1908 (see also Jeffrey 1962; Keddy 1981) to explain the trend towards diploid growth in land plants. It is based on the notion that gametes and spores have different and opposed requirement for dispersal. Haploid meiospores that will develop into independent freeliving gametophytes need to be dispersed away from the parent plant to new sites suitable for growth. Dispersal in air is more effective when the spores are released from a greater height: the sporophyte should therefore be a tall erect plant. Even in bryophytes, the sporophyte is usually a conspicuously erect structure growing above the leafy gametophyte. Motile haploid gametes that can fuse only if they encounter a suitable partner should be aggregated, not dispersed, in conditions where free water is available for movement: the gametophyte should therefore be a low or subaerial thallus. In the largest plants, of course, the gametophyte is still further specialized and reduced, as a consequence of the ability of the male gametophyte to grow through the tissues of the sporophyte towards the sessile female gametophyte. This theory offers a very clear account of evolutionary trends in plants, although it has never been extended to other groups of organisms. It is not clear whether it can explain why some unicellular groups are haplontic and others diplontic; or why sporophytes and gametophytes differ in form among aquatic groups where the viscosity of the medium puts less of a premium on height as a prerequisite for spore dispersal. Nevertheless, the differing requirements of haplospores for aggregation and dispersal offers in principle a framework for interpreting the diversity of life cycles that should be taken more seriously by theoreticians.

Sexual selection and natural selection. The clearest genetic consequence of the alternation of generations is that haploid growth yields uniform gametes through mitosis, whereas diploid growth yields diverse gametes through meiosis (Bell 1982). Diploid growth will therefore be favoured when there is competition among gametes for fusion partners, because diploid individuals will produce a small proportion of highly superior gametes. Haploid growth will be favoured when competition is minimal, since the uniform gametes produced by a haplont will almost all be functional, and will outcompete the large fraction of incompetent gametes produced by meiotic recombination in diplonts. The flaw in this hypothesis is that it is difficult to explain how genetic variance for gamete performance is maintained; genes which improve gamete performance should quickly become fixed in the population, removing the genetic basis for gamete competition and therefore for effective selection.

I shall now suggest that some loci will affect both gamete performance and vegetative growth and reproduction. This will lead to a tension between sexual selection, acting through gamete performance, and natural selection, acting through growth and reproduction. Alleles at any one of these loci may increase performance, whether expressed in the gamete or in the vegetative individual; such alleles will quickly be fixed. Other alleles will decrease performance both in the sexual and in the vegetative cycles; these will be rapidly eliminated. The alleles which remain segregating in the population for substantial periods of time will be those which have opposed effects on fitness in the sexual and vegetative cycles-- increasing gamete performance but decreasing growth, or vice versa. Diploid growth will be favoured when this sort of antagonistic pleiotropy exists, because the long-term maintenance of genetic variance for gamete performance permits gamete competition to be renewed in every generation. Diploid growth should therefore evolve when there is negative genetic correlation between the effects of genes on sexual and vegetative function, which is likely to be the case when the gametes are highly specialized. Conversely, haploid growth will be favoured by positive genetic correlation, which is likely to be the case when gametes are similar to vegetative cells.

The simplest formal representation of this theory is a two-locus model in which one locus controls the relationship between the sexual and vegetative cycles, while a second, unlinked locus affects sexual and vegetative performance. The model was given verbally by Bell (1982). Formal two-locus models for viability selection have been developed by Jenkins (1993) and used to derive conditions for the duration of haploid and diploid growth in the life cycle. The two alleles at the life cycle locus are $\underline{M}$, directing a diplontic cycle with gametic meiosis, and $\underline{m}$, directing a haplontic cycle with a zygotic meiosis. These genes act in the zygote, such that the first division of the zygote is either mitotic ($\underline{M}$) or meiotic ($\underline{m}$); either may be dominant. The two alleles at the fitness locus have the following effects:

	Performance:	
Genotype	Gametic	Vegetative
A	1-v	1
a	1	1-w
AA		1
Aa		1-hw
aa		1-w

The parameters v and w are treated most simply as if they were survival rates; h represents the degree of dominance of A in the sporophyte. Mating is at random, so that the genes which switch the life cycle to haplonty or to diplonty are segregating in the same Mendelian population. The verbal argument in the preceding paragraph leads one to expect that diplonty will evolve when there is negative genetic correlation between sexual and vegetative function, or that M will increase in frequency if v and w have the same sign.

The fitness (A,a) locus goes to fixation in one direction or the other. As it does so, it induces a change in frequency at the life-cycle (M,m) locus; once the fitness locus is fixed, there is no further change in gene frequency at the life cycle locus. The shift in frequency of the life-cycle genes occurs because they become correlated under selection with genes at the fitness locus, so that selection acting directly on the fitness locus causes indirectly changes in frequency at the life-cycle locus. The correlation between genes at the two loci can be represented by the coefficient of linkage disequilibrium, $D = f_{MA}f_{ma} - f_{Ma}f_{mA}$; if D is forcibly reset to zero at the beginning of each generation then any correlation which selection has generated between the two loci is destroyed, and as the fitness locus goes to fixation in one direction or the other there is no change in frequency at the life-cycle locus. We wish to know whether M or m increase in frequency during selection at the fitness locus. The full behaviour of the model is somewhat complicated, in part because of the interaction between dominance at the two loci. However, analytical solutions are readily obtained for some special cases. In particular, if there is weak selection (v and w small in absolute magnitude) and no dominance at the fitness locus ($h = \frac{1}{2}$), then the necessary and sufficient condition for M to increase in frequency is

(1) $$D (v - w/2) < 0 \quad \text{(Figure 5)}.$$

Selection creates a correlation between the m gene, directing haplonty, and the gene which has the better vegetative performance. Thus, if $w > 0$ then $D < 0$, because of the prevalence of mA and Ma chromosomes. Diplonty then increases if the advantage of the Ma chromosome in gametes more than balances its disadvantage in vegetative individuals, which for the case of no dominance is true if $v > w/2$. The holds regardless of dominance at the life-cycle locus, and iterating the recursions shows that it is approximately true for strong selection. The advantage of diplonty can be thought of as lying in the ability of the Aa heterozygote to be vegetatively successful and at the same time to produce superior gametes. If $v > 0$ and $w > 0$ then the a gamete is superior, whereas A is vegetatively superior; the

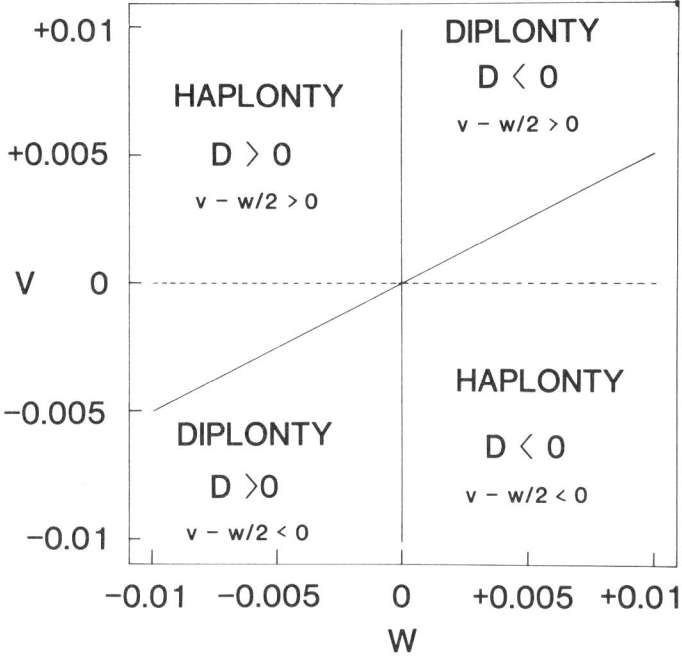

FIGURE 5. Evolution of haplontic or diplontic cycles as the outcome of antagonism between sexual selection and natural selection. The model is described in the text.

Aa heterozygote (provided that $h < 1$) can therefore grow fairly well by virtue of the partial expression of its A gene, and at the same time produce a gametes. An antagonism between the effects of A and a tends to retard fixation at the fitness locus; if genes with these properties are maintained in the population permanently then diplonty will become fixed. Conversely, when A and a have similar effects on sexual and vegetative function then haplonty is favoured.

A very simple model is, then, capable of showing that either haploid or diploid growth may be favoured, depending on how sexual and vegetative competition are related.

The theory requires that many loci are expressed in both the sexual and the vegetative cycles. This seems likely to be the case; it has suggested that as much as 60% of the genome of Arabidopsis is expressed in both sporophyte and gametophyte, and in plants gametophytes compete for fusion rather than for growth. It has also been suggested that gametophytic (fertilization) and sporophytic (growth) performance are positively correlated in plants (see Mulcahy & Mulcahy 1975). However, although this may be true among unselected mutants and recent isolates I doubt that it is true at equilibrium, for the reasons I have given above. The appropriate experiment is to select upwards for fertilization success, expecting a correlated reduction in vegetative growth rate, and vice versa. Da Silva & Bell (1992) obtained the expected result in Chlamydomonas: selection for mating efficiency reduced vegetative growth-rate, while selection for vegetative growth reduced mating efficiency. This shows that genes with the required properties are present and respond to selection, even in a unicellular haplont.

The main comparative prediction of the theory is that there should be correlated trends in diploid growth, vegetative size and complexity, and gamete specialization. The very brief comparative account I have had space for shows (as has long been known) that such trends exist. Unicellular eukaryotes with unspecialized gametes resembling vegetative cells or zoospores are haplontic, and most are isogametic. This rule is not followed by forms with gametangial or gamontic copulation, where gamete competition is minimized. Diploid growth generally characterizes larger multicellular forms with specialized and differentiated gametes, and the same trend exists within clades such as phaeophytes, chlorophytes and green plants.There are many exceptions to these broad general trends, and a great deal of unexplained variation among life cycles. However, the interpretation of the alternation of generations that I have favoured here, in terms of the relationship between the sexual and vegetative cycles, shows how detailed, case-by-case interpretations can be built up from a knowledge of fertilization and growth. Developing and testing these interpretations is the next stage in a synthetic evolutionary theory of life cycles.

REFERENCES

Baroin, A., Perasso, R., Qu, L.-H., Brugerolle, G., Bachellerie, J.-P. & Adoutte, A. Partial phylogeny of the unicellular eukaryotes based on rapid sequencing of a portion of 28S ribosomal RNA. Proc. Nat. Acad. Sci. USA **85** (1988), 3474-3478.

Bell, G. The Masterpiece of Nature. Croom Helm, London; The University of California Press, Berkeley, 1982.

Bell, G. Sex and Death in Protozoa. Cambridge University Press, Cambridge, 1989.

Bell, G. & Koufopanou, V. The architecture of the life cycle in small organisms. Phil. Trans. R. Soc. London B **332** (1991), 81-89.

Bold, H.C. & Wynne, M.J. Introduction to the Algae. Prentice-Hall, Inglewood Cliffs, New Jersey, 1978.

Bower, F.O. The Origin of a Land Flora. Macmillan, London, 1908.

Cavalier Smith, T. (ed.) The Evolution of Genome Size. Wiley, Chichester, New Hampshire, 1985.

Crow, J.F. & Kimura, M. Evolution in sexual and asexual populations. Amer. Natur. **99** (1965), 439-449.

Da Silva, J. & Bell, G. The ecology and genetics of fitness in Chlamydomonas. VI. Antagonism between natural selection and sexual selection. Proc. R. Soc. London B **249** (1992), 227-233.

Devereux, R., Loeblich, A.R. & Fox, G.E. Higher plant origins and the phylogeny of green algae. J. Mol. Evol. **31** (1990), 18-24.

Douglas, S., Murphy, C., Spencer, D. & Gray, M. Cryptomonad algae are evolutionary chimaeras of two phylogenetically distinct unicellular eukaryotes. Nature **350** (1991), 148-151.

Elwood, H.J., Olsen, G.J. & Sogin, M.L. The small-subunit ribosomal RNA gene sequences from the hypotrichous ciliates Oxytricha nova and Stylonichia pustulata. Mol. Biol. Evol. **2** (1985), 399-410.

Hendriks, L., De Baere, R., Van de Peer, Y., Neefs, J., Goris, A. & De Wachter, R. The evolutionary position of the rhodophyte Porphyra umbilicalis and the basidiomycete Leucosporidium scoltii among other eucaryotes as deduced from complete sequences of small ribosomal subunit RNA. J. Mol. Evol. **32** (1991), 167-177.

Hofmeister, W. Vergleichende Untersuchungen der Keimung, Entfaltung und Fruchtbildung hoherer Kryptogamen und der Samenbildung der Coniferen. Leipzig, 1851. [English trans. The Ray Society, 1862.]

Jeffrey, C. The origin and differentiation of the archegoniate land plants. Bot. Not. **115** (1962), 446-454.

Jenkins, C.D. Selection and the evolution of genetic life cycles. Genetics **133** (1993), 401-410.

Keddy, P.A. Why gametophytes and sporophytes are different: form and function in a terrestrial environment. Amer. Natur. **118** (1981), 452-454.

Kondrashov, A.S. & Crow, J.F. Haploidy or diploidy: which is better? Nature **351** (1991), 314-315.

Lenaers, G., Maroteaux, L., Michot, B. & Herzog, M. Dinoflagellates in evolution: a molecular phylogenetic analysis of large subunit ribosomal RNA. J. Mol. Evol. **29** (1989), 40-51.

Lenaers, G., Scholin, C., Bhaud, Y., Saint-Hilaire, D. & Herzog, M. A molecular phylogeny of dinoflagellate protists (Pyrrhophyta) inferred from the sequence of 24S rRNA divergent domains D1 and D8. J. Mol. Evol. **32** (1991), 53-63.

Lewis, J. & Wolpert, L. Diploidy, evolution and sex. J. theoret. Biol. **78** (1979), 425-438.

Lewis, W.M. Nutrient scarcity as an evolutionary cause of haploidy. Amer. Natur. **125** (1985), 692-701.

Lobban, C.S. & Wynne, M.J. (eds.) The Biology of Seaweeds. Blackwell Scientific Publications, Oxford, 1981.

Margulis, L., Corliss, J.O., Melkonian, M. & Chapman, D.J. Handbook of Protoctista. Jones & Bartlett, Boston, 1990.

Martin, W., Somerville, C.C. & Loiseaux-de Goer, S. Molecular phylogenies of plastid origins and algal evolution. J. Mol. Evol. **35** (1992), 385-404.

Maynard Smith, J. The Evolution of Sex. Cambridge University Press, Cambridge 1978.

Mortlock, R.P. (ed.) Microorganisms as Model Systems for Studying Evolution. Plenum Press, New York, 1984.

Mulcahy, D.L. & Mulcahy, G.B. The influence of gametophytic competition on sporophytic quality. Theoret. Appl. Genet. **46** (1975), 277-280.

Muller, H.J. Some genetic aspects of sex. Amer. Natur. **8** (1932), 118-138.

Paquin, C. & Adams, J. Frequency of fixation of adaptive mutations is higher in evolving diploid than haploid yeast populations. Nature **302** (1983), 495-500.

Perasso, R., Baroin, A., Qu, L.H., Bachellerie, J.P. & Adoutte, A. Origin of the algae. Nature **339** (1989), 142-144.

Perrot, V., Richerd, S. & Valero, M. Transition from haploidy to diploidy. Nature **351** (1991), 315-317.

Qu, L.-H., Perasso, R., Baroin, A., Brugerolle, G., Bachellerie, J.-P. & Adoutte, A. Molecular evolution of the 5'-terminal domain of large-subunit rRNA from lower eukaryotes. A broad phylogeny covering photosynthetic and non-photosynthetic protists. BioSystems **21** (1988), 203-208.

Sogin, M.L. & Elwood, H.J. Primary structure of the Paramecium aurelia small-subunit rRNA coding region: phylogenetic relationships within the Ciliophora. J. Mol. Evol. **23** (1986), 53-60.

Steenstrup, J.J. On the Alternation of Generations: or, the Propagation and Development of Animals through Alternate Generations. English trans. Busk, G., The Ray Society, London, 1845.

Strasburger, E. The periodic reduction in the number of chromosomes in the life-history of living organisms. Ann. Bot. **8** (1894), 295.

Tanner, C.E. Chlorophyta: life histories. Pp 133-193 in Lobban, C.S. & Wynne, M.J. (eds.) The Biology of Seaweeds. Blackwell Scientific Publications, Oxford, 1981.

Wainwright, P.O., Hinkle, G., Sogin, M.L. & Stickel, S.K. Monophyletic origins of the Metazoa: an evolutionary link with fungi. Science **260** (1993), 340-342.

DEPARTMENT OF BIOLOGY, MCGILL UNIVERSITY, 1205 AVENUE DOCTEUR PENFIELD,

MONTREAL, QUEBEC H3A 1B1, CANADA

Lectures on Mathematics in the Life Sciences
Volume **25**, 1994

Gradual Origin of Amphimixis by Natural Selection

A. S. KONDRASHOV

ABSTRACT. Modern amphimixis, alternation of syngamy and meiosis, differs
from apomixis, the production of offspring only by mitosis, by three principal
features: syngamy, reduction, and recombination. In this paper I construct a
plausible scenario for the evolution of amphimixis from apomixis, such that
every intermediate step is favored by natural selection. The steps are: 1)
establishment of the ploidy cycle consisting of syngamy and primitive
reduction, 2) regularization of reduction, and 3) origin of crossing-over.
Possible cytological mechanisms of each step are discussed. Epistatic selection
against slightly deleterious mutations can favor step 1) if the ancestral apomict
had more than one chromosome because independent assortment of non-
homologous chromosomes still can lead to a significant increase of the variance
in the per genome number of mutations when there is no crossing-over. Even if
the primitive ploidy cycle resulted in frequent aneuploidy, it could be favored
as a facultative mode of reproduction if intensity of selection against mutations
fluctuated from generation to generation. The absence of crossing-over can
facilitate the invasion of the modifier that causes amphimixis into the apomictic
population. Step 2) could lead to a large and immediate advantage, abolishing
the cost of aneuploidy. If the ancestral apomict had an "endomitotic ploidy
cycle" due to alternation of endomitosis and reduction, regular reduction may
predate syngamy, making step 2) unnecessary. Under synergistic epistasis,
crossing-over always reduces the mutation load and increases the variance of
the number of mutations per genome. However, individual selection favors step
3) only when selection against mutations is strong.

> " ... adoption of sexual reproduction ... depended on the invention of meiosis
> as a modification of mitosis. The fusion of cells and nuclei is such a
> commonplace accident ... that the introduction of fertilization involves no
> novelty. Meiosis on the other hand is an abrupt and revolutionary change
> which permits of no half-way house. The chromosomes must either reduce or
> fail to reduce if they are to keep their genetic character. Anything
> intermediate upsets the whole apple-cart. ... The origin of meiosis and sexual

1991 Mathematics Subject Classification: 92D15

I am grateful to M. Maguire for useful criticism and to R. Harrison, D. Houle, and B.
Normark who made many comments on the manuscript.

This paper is in final form and no version of it will be submitted for publication elswhere.

27

reproduction therefore shows the most violent discontinuity in the whole of evolution. ... It is impossible to imagine it as the result of a gradual accumulation of changes each one of which had a value as an adaptation ... . If the material processes underlying sexual reproduction had been understood at the time neither Lamarck nor Darwin could ever have thought of evolution as depending on the adaptive accumulation of entirely continuous variation.

C. D. Darlington, 1939, p. 125

1. Introduction

The goal of this article is to dispute Darlington's conclusion. The life cycles of the majority of contemporary eukaryotes include amphimixis (sexual reproduction), which consists of the alternation of syngamy (gamete fusion, fertilization) and reduction, dividing the life cycle into haploid and diploid phases. Of course, mitosis, either in haploid, diploid, or in both phases, is also a necessary part of amphimictic life cycles. However, obligate apomixis, under which an offspring always appears from a single mitotically produced cell, so that the genes from different organisms never come together, is relatively rare (see Margulis et al., 1990; Bell, 1993).

Recent interest in the evolution of reproduction by natural selection has concentrated on two problems: 1) competition between obligately amphimictic and apomictic populations and 2) evolution of the features of amphimixis such as recombination and outcrossing (see Kondrashov, 1993, for a review). Of course, in both cases amphimixis must be already present. Therefore, the origin of amphimixis has not attracted much attention. Still, contrary to the scepticism of Darlington (1939), several cytological scenarios of a gradual origin of amphimixis from apomixis were proposed (Cleveland, 1947; Raikov, 1982; Margulis and Sagan, 1986; Maguire, 1992). However, the question "can natural selection favor all the postulated intermediate steps?" has either been ignored or not answered satisfactorily. Several population models may be relevant (Kondrashov, 1985; Uyenoyama and Bengtsson, 1989; Bengtsson 1992), but they all imply that from the very beginning amphimictic reduction occured by the modern meiosis, thus ignoring the main problem in the origin of amphimixis.

Almost certainly amphimixis appeared from apomixis in some unicellular eukaryote (or eukaryotes) which already had mitosis. The majority of contemporary eukaryotic taxa contain at least some amphimicts. However, several groups of unicellular eukaryotes seem to lack amphimixis completely, although in some cases it may be overlooked (see Margulis et al., 1990). Multicellular eukaryotes are generally amphimictic, and multicellular apomicts are of secondary origin. In the case of unicellular forms, however, it is very difficult to decide whether the absence of amphimixis is a primitive or a secondary condition. Depending on this, amphimixis may be either polyphyletic (Ivanov, 1970; Raikov, 1982, pp. 206-209), or its origin may be a unique event (Heywood and Magee, 1976).

Here the origin of amphimixis will be investigated, considering both the hypothetical intermediate steps and selection pressures that could favor them. I will start from a discussion of the possible cytological mechanisms of these steps, but my main purpose is to understand the evolution of gene transmission from parents to offspring, and not the underlying changes at the cellular level.

2. Transition from Apomixis to Amphimixis: Cytology

2.1. Overview. In modern amphimixis, alternation of syngamy and reduction is almost always strict, which leads to regular alternation of diploid and haploid phases, at least in the germ line, although endomitosis can lead to additional phases in a brown alga *Ectocarpus siliculosus* (Muller, 1967) and perhaps in some red algae (see Haig, 1993). Reduction is almost always regular and achieved by a two-step meiosis (consisting of DNA replication followed by two successive cell divisions). Meiosis almost always involves crossing-over. Certainly, primitive amphimixis was very different.

The origin of amphimixis was marked by the origin of regular syngamy, which combines genes from different organisms into one cell, and eventually into one nucleus. Syngamy is probably the only part of amphimixis whose genetic consequences have remained unchanged from the very beginning. As a regular (even very rare) event, syngamy is impossible without reduction, the two-fold decrease of ploidy, because many successive syngamies would lead to an impossible level of ploidy. Thus, reduction either must have appeared almost immediately after syngamy became a regular phenomenon or could have existed even before it, if the ancestral apomict had an endomitotic ploidy cycle (see below). Of course, one or several syngamies may accidentally occur in a lineage incapable of reduction, but this would lead to permanently di- or polyploid apomicts, not to the origin of amphimixis. Thus, the conclusion seems unavoidable that the first step in the evolution of amphimixis was establishment of a ploidy cycle which included syngamy and reduction.

After this step, the primitive form of amphimixis was probably different from the modern one in the structure of the ploidy cycle, in the mechanism of reduction, and in the mode of recombination. The primitive ploidy cycle could be irregular, with two or more successive syngamies followed by reductions, thus leading to cycles with haploid, diploid, and polyploid phases. The possibility (never observed in nature) that, say, three gametes could take part in syngamy followed by three-fold reduction, will be ignored. Primitive reduction may have also been irregular, i. e. frequently resulted in aneuploid products. Primitive recombination may have consisted only of independent assortment of non-homologous chromosomes during reduction without any crossing-over, provided that there were more than one chromosome in the haploid genome. It is difficult to imagine how reduction and crossing-over can appear simultaneously, or why crossing-over should evolve before reduction. Therefore, the possibility of their separate evolution is very important.

Thus, the problem of the origin of amphimixis can be partitioned into three subproblems:

    a) What was the ancestral apomict?

    b) How did the primitive amphimictic ploidy cycle evolve?

    c) How did the modern amphimixis evolve from a primitive one?

2.2. Ancestral apomict. There are two broad possibilities regarding the ancestral apomict: 1) it reproduced only by ordinary mitosis, or 2) it had a more complicated "endomitotic" ploidy cycle (Cleveland, 1947; Hurst and Nurse, 1991). Subsequent evolution of amphimixis could be different in these two cases. In both cases I will assume that the ancestral apomict had more than one chromosome in the haploid genome, because otherwise I do not see how natural selection could favor primitive amphimixis (see below).

Actually, even mitosis already involves a kind of ploidy cycle: after replication and before cell division each chromosome is represented by two chromatids, instead of one before replication. However, in contrast to the "real" ploidy cycles (endomitotic and amphimictic), during mitosis a cell division is always preceeded by replication, so that reduction does not occur.

An endomitotic ploidy cycle occurs in apomicts if sometimes replication is not followed by cell division (endomitosis; similar processes create polytene chromosomes in some amphimicts). Cell divisions without previous replication must also occur, to prevent unlimited increase of ploidy. The result is the alternation of haploid and diploid phases or, if several endomitoses are followed by several reductions, a cycle with haploid, diploid, and polyploid phases. Endomitotic ploidy cycles ("cyclic polyploidy" according to Raikov, 1982, p. 146) are known in some contempotrary eukaryotes (see Discussion). Reduction in these cycles appears to be regular. Thus, if amphimixis originated from apomicts with an endomitotic ploidy cycle, a mechanism of regular reduction may already have been available when syngamy appeared for the first time.

During an endomitotic ploidy cycle the cells at some or all levels of ploidy may also reproduce by mitosis. The pre-existence of di- or polyploid mitosis might be another feature of this cycle facilitating the origin of amphimixis. This may not be important, however, if initially either 1) the diploid phase was very brief so that reduction immediately followed syngamy, without diploid mitoses in between or 2) mitosis under a higher level of ploidy does not require any new mechanisms (as the data on modern polyploids suggest).

2.3. Origin of the amphimictic ploidy cycle. The nature of the primitive reduction process which appeared immediately after the origin of regular syngamy is the most puzzling feature of the origin of amphimixis. Two broad possibilities are irregular and regular reduction. For simplicity I assume that syngamy and reduction are always strictly alternating. This probably was not the case, but the consequences of such an assumption do not seem too important.

It is difficult to imagine how amphimictic reduction could be regular from the very beginning if the ancestral apomict did not have the endomitotic ploidy cycle. A possible mechanism of the primitive irregular reduction is random loss of chromosomes, caused by a series of mitotic divisions with nondisjunctions, and similar to parasexual process in some modern fungi (Maguire, 1992).

Any irregular reduction is costly, because it produces many inviable aneuploids. If reduction occurs by random loss of individual chromosomes, the probability that a euploid haploid combination will be reached starting from a diploid cell is

$$2^n \binom{2n}{n}^{-1}$$, where n is a haploid chromosome number (2^n out of $\binom{2n}{n}$ possible combinations of n chromosomes are euploid). If, alternatively, a diploid cell divides and chromosomes migrate to one or the other daughter cell independently, the expected number of euploid products of such division is $2^{-(n-1)}$, because there are two daughter cells and segregation of each chromosome pair leads to euploid combination with the probability 0.5 (the related calculations of Margulis and Sagan (1986, Table 11) are wrong). We can see (Fig. 1) that in both cases the probability of producing a euploid combination is substantial if n < 5-7, but rapidly approaches zero when n becomes larger.

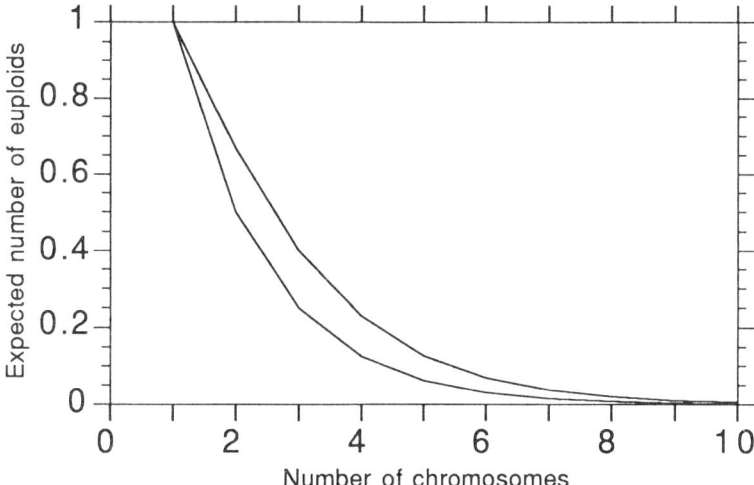

FIGURE 1. Expected number of euploid products after irregular reduction by random chromosome loss (upper line) and by random distrbution of chromosomes between two daughter cells (lower line).

In contrast, if the ancestral apomict had the endomitotic ploidy cycle, syngamy may simply substitute for endomitosis as the means of DNA doubling, after which regular reduction can occur due to a pre-existing mechanism. However, even if the ancestral apomict had such a cycle, this does not guarantee that primitive amphimictic reduction was regular. Reduction after syngamy involves homologous chromosomes which come from different cells and thus are not, at least initially, attached to each other. Thus, if after endomitosis homologous chromosomes never become completely separated, an endomitotic ploidy cycle may not be sufficient to ensure regularity of primitive amphimicitc reduction. In contrast, if during this cycle homologous chromosomes are separated before reduction, then reduction after syngamy can be identical to reduction after endomitosis. In this case regular reduction can involve re-establishing (endomitotic cycle) or establishing (amphimictic cycle) the contact between homologous chromosomes, or can go completely without it.

Regular primitive reduction with contact between homologous chromosomes may be similar to modern one-step meiosis, during which homologous unreplicated chromosomes in a diploid cell form pairs and then go into different haploid daughter cells. One-step meiosis was reported in *Pyrsonymphida*, *Parabasalia*, *Dinoflagellata*, *Apicomplexa*, *Microspora*, and *Myxozoa* (see Raikov, 1982, pp. 202-209; Margulis et al., 1990), although these data have recently been disputed (Haig, 1993). Similar to the ordinary two-step meiosis, one-step meiosis does not normally lead to aneuploidy and differs from ordinary mitosis in a haploid cell only by the origin of "chromatid pairs" that are split after metaphase. In mitosis each pair consists of identical sister chromatids, created during pre-mitotic DNA replication, and retaining some physical contact. In contrast, one-step meiosis includes pairing of initially separate homologous chromosomes, and "chromatids" are, in fact, different chromosomes.

Evolution of one-step meiosis from mitosis can be easy, provided that homologous chromosomes can find each other. "If we suppose that the ordinary replication of chromosomes is omitted before one of the divisions of the diploid nucleus in which homologous chromosomes are somatically paired, homologous chromosomes instead of sister chromatids will part into the daughter nuclei. The result is one-step meiosis" (Raikov, 1982, p. 207).

Alternatively, a regular reduction can occur without any physical contact between homologous chromosomes (Oakley, 1985; John, 1990, p. 221). Currently this phenomenon is of secondary origin, but its existence indicates the feasibility of such a process. Almost certainly, the primitive amphimictic reduction, even if it was regular, was different from modern two-step meiosis, which can hardly appear as a part of an endomitotic ploidy cycle.

Even if primitive segregation required some form of contact between homologous chromosomes, chiasmata were not necessary for this purpose, as is indicated by the cases of one-step (Raikov, 1982) and two-step (Jones, 1990, p. 86) achiasmatic meiosis in modern organisms (although achiasmatic segregation in modern two-step meiosis is of secondary origin). Thus, crossing-over was not a necessary part of the primitive amphimictic reduction and could evolve much later.

2.4. Regularization of reduction. If the primitive amphimictic reduction was irregular, selection obviously strongly favored its regularization. Reduction could have become regular before homologous chromosomes had begun to pair. Alternatively, if regularization was involved in the origin of chromosome pairing, the resulting process could have been similar to either one-step or two-step meiosis. In the first case one-step meiosis observed now may be a primitive condition, while in the second case it is a secondary derivative of two-step meiosis. Molecular mechanisms that allow homologous chromosomes to pair have probably evolved from the mitotic pathway for the repair of double-strand DNA breaks (Kleckner et al., 1991; McKee et al., 1992).

Origin of one-step meiosis from irregular reduction by random loss of chromosomes may be similar to its origin from mitosis and again requires only suppression of premeiotic replication and pairing of homologs. Later, two-step meiosis could evolve, due to addition of premeiotic replication and the second cell division. Alternatively, two-step meiosis could evolve directly from irregular reduction, by adding the pairing of homologs and the second meiotic division (Maguire, 1992).

2.5. Origin of crossing-over. If reduction does not follow syngamy immediately, crossing-over, in principle, can occur at any moment in the diplophase: immediately after syngamy, somethime during diploid interphase or diploid mitosis, or during reduction. Genetical consequences of all these processes to the genotype distribution after the return to haploidy would be the same. However, although crossing-over sometimes occurs in mitosis, currently the regular high-frequency crossing-over occurs only during meiosis, so that gametic chromosomes remain intact until the very end of the diplophase. One possible explanation for this is that a close contact between homologous chromosomes, which is necessary for reciprocal crossing-over, had evolved as a part of reduction to ensure regular segregation. This could make reduction the most suitable platform for the origin of crossing-over.

High-frequency crossing-over is known only from the first division of two-step meiosis, while in all the reported cases of one-step meiosis it is apparently absent (Raikov, 1982). It is unclear why this is the case, because each instance of crossing-over apparently involves only two, and not four, DNA molecules. Indeed, mitotic crossing-over may occur before DNA replication (see Jones, 1990, p. 194).

Whatever the details of the origin of crossing-over might be, it probably evolved after the origin of regular reduction and chromosome pairing. The synaptonemal complex, which is now required for precise alignment of homologs and reciprocal crossing-over "may have evolved in meiosis at a second stage, subsequent to the basic pairing mechanism" (Kleckner et al., 1991). This supports the view (Margulis and Sagan, 1986) that the origin of regular high-frequency crossing-over was the separate and perhaps the last event in the origin of amphimixis (for the opposite view see Penny, 1985).

3. Transition from Apomixis to Amphimixis: Selection

3.1. Overview. If many unrelated events must occur before a phenotype becomes beneficial, it is unlikely to evolve by natural selection. Thus, because an instant creation of modern amphimixis from apomixis seems impossible, the origin of amphimixis must have been gradual, and during the accumulation of genetic changes affecting the mode of reproduction, every intermediate mode must have been favored by selection. Instead of postulating any immediate benefits of syngamy, reduction, or crossing-over, which is not supported by the data (see Kondrashov, 1993), I will assume that the factor that governed the origin of amphimixis (except step 2)) was epistatic selection against slightly deleterious mutations.

3.2. Origin of the amphimictic ploidy cycle. Consider the ancestral apomictic population in mutation-selection equilibrium. Assume for simplicity that the population was haploid and that selection was truncation (genotypes with T or less mutations have fitness 1, while all others die). Then before selection all individuals carried exactly T mutations, and the mutation load L was $1-e^{-U}$, where U is the genomic deleterious mutation rate (in an apomictic population under a given U the load is the same under any mode of selection, Kimura and Maruyama, 1966).

In contrast, amphimictic populations are known to maintain some variance in the number of mutations in the genomes even with truncation or other forms of epistatic selection, which reduces the mutation load under such selection (see Kondrashov, 1993). This was demonstrated for amphimictic populations with either free recombination (Kimura and Maruyama, 1966; Crow, 1970; Kondrashov, 1982) or with a limited, but non-zero, rate of crossing-over (Kondrashov, 1984; Charlesworth, 1990). The case of n > 1 was previously investigated only using a Monte-Carlo model (Kondrashov, 1984). Here it will be studied by deterministic iterations, assuming that n = 2 and crossing-over is absent, so that recombination is caused only by independent assortment of non-homologous chromosomes.

Consider a population with the following life cycle: selection - reproduction (consisting of syngamy followed by reduction, so that selection acts in the haplophase) - mutation. Suppose that n = 2. Then, if all mutant alleles are equally deleterious, the population can be described by the distribution p(i,j), the frequency of individuals with i mutations in the first and j mutations in the second chromosome. The equations which describe transformations of p(i,j) during all the processes of the life cycle are a straightforward extension of eq. 3.1 from Kimura and Maruyama (1966). Numerical solution of these equations allows us to find the properties of equilibrium populations. The results, together with the results for no recombination and free recombination, are presented in Fig. 2. All the programs are written in THINK C for Macintosh and are available on request.

We can see that the mutation load in the population with n = 2 and without crossing-over (I assumed that both chromosomes are of equal size, so that the mutation rate for each of them is U/2), is much lower than that under apomixis, although free recombination reduces the load even further (Fig. 2a). Even without crossing-over, amphimixis maintains a substantial equilibrium variance in the number of mutations in the genome, while under apomixis the variance equals zero and all individuals carry exactly T mutations (Fig. 2b).

Actually, the mutation load in the case of n = 2 and no crossing-over is only slightly higher than $1 - e^{-U/2}$. The reason for this is the following. According to Kimura and Maruyama (1966), the mean fitness of individuals which have no mutations in one of the two chromosomes must be exactly $e^{-U/2}$ times higher that the mean population fitness. With only two chromosomes such individuals have a substantially diminished genomic number of mutations, and their mean fitness is only slightly lower than one. When n grows, the load in a population without crossing-over exeeds $1 - e^{-U/n}$ more and more.

Thus, even recombination caused exclusively by independent assortment of only two chromosomes can have significant consequences. Under large U the decrease of the load can be sufficient to offset even a high cost of aneuploidy. Even if the total number of mutations in all parents was exactly T at some initial moment, different chromosomes would have carried different numbers of mutations, and this variance could have been released by recombination due to independent assortment of chromosomes.

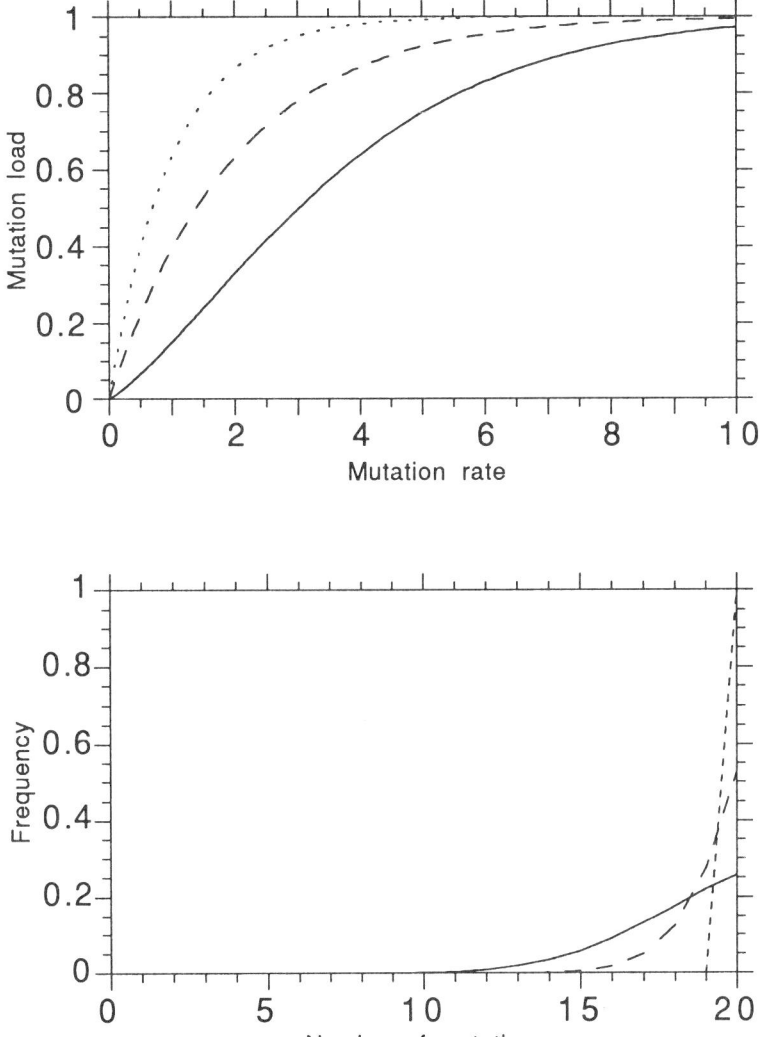

FIGURE 2. a) Mutation load as a function of U under truncation selection (T = 20) with apomixis, amphimixis with n = 2 and no crossing-over, and amphimixis with free recombination (dotted, broken, and and solid line, respectively). b) Equilibrium distributions of the per genome number of mutations after selection with these 3 modes of reproduction under U = 4.0 and the same selection.

However, this consideration involves only group selection, which is appropriate if we compare the features of obligately amphimictic and obligately apomictic populations, but is not sufficient to study the origin of amphimixis. Thus, individual selection must be investigated.

For this purpose, consider a modifier locus with two alleles, s and S. Individuals carrying the allele s do not initiate syngamy, and, thus, when only this allele is present, the population is obligately apomictic. Individuals with allele S initiate syngamy with a randomly chosen individual (either s or S) with probability P (penetrance of S), and do not initiate syngamy with probability 1-P. Syngamy is immediately followed by reduction, which produces two viable euploid cells with the probability F. Reduction can lead to either free recombination (in this case the number of chromosomes does not matter) or independent assortment of non-homologous chromosomes without crossing-over. In this case, as before, n = 2 and both chromosomes have the same size, while the modifier locus is situated either on one of them or on a tiny third chromosome that carries no other genes.

If recombination is free, the total number of mutations in both chromosomes, together with a genotype at the modifier locus, is sufficient to describe the organism, and the equations describing transformations of the population during the life cycle can be found in Kondrashov (1985). If crossing-over is absent, the population is completely described by the distribution $q(k,i,j)$, the frequency of individuals of the k-th genotype at the modifier locus (s and S correspond to k = 0 and 1, respectively) having i and j mutations in the first and the second chromosomes, respectively. The corresponding equations are an extension of eqs. 1-3 from Kondrashov (1985).

If the fitness of a genotype with i mutations is $w(i)$, the distribution after selection, $q'(k,i,j)$ can be obtained from that before selection, $q(k,i,j)$ by:

$$(1) \quad q'(k,i,j) = q(k,i,j)w(i+j)/\overline{W}; \quad \text{where} \quad \overline{W} = \sum_{i=0}^{T} \sum_{j=0}^{T-i} \sum_{k=0}^{1} q(k,i,j)w(i+j)$$

During reproduction a fraction 1-xP of individuals with the allele s, and fraction (1-P)(1-xP) of individuals with the allele S reproduce apomictically, where x is the frequency of S:

$$(2) \quad x = \sum_{i=0}^{T} \sum_{j=0}^{T-i} q'(1,i,j)$$

Individuals with the allele s that reproduce amphimictically always mate with individuals with the allele S, while the individuals with the allele S that reproduce amphimictically mate with individuals with alleles s and S with the probabilities P(1-x)/[P(1-x+xP)] and [Px(2-P)]/[P(1-x+xP)], respectively. I assume that mating is random with respect to the numbers of mutations. Then in the case when the modifier is unlinked to both chromosomes, reproduction leads to the following changes of q:

$$q''(0,i,j) = [(1 - Px)q'(0,i,j) +$$

$$\frac{1}{4}FP \sum_{i_1=0}^{T} \sum_{j_1=0}^{T-i_1} \sum_{i_2=0}^{T} \sum_{j_2=0}^{T-i_2} q'(0,i_1,j_1)q'(1,i_2,j_2)(\delta_{i_1}^i + \delta_{i_2}^i)(\delta_{j_1}^j + \delta_{j_2}^j)]/D$$

(3a)

$$q''(1,i,j) = [(1-Px)(1-P)q'(1,i,j) +$$

$$\frac{1}{4}FP \sum_{i_1=0}^{T} \sum_{j_1=0}^{T-i_1} \sum_{i_2=0}^{T} \sum_{j_2=0}^{T-i_2} q'(0,i_1,j_1)q'(1,i_2,j_2)(\delta_{i_1}^i + \delta_{i_2}^i)(\delta_{j_1}^j + \delta_{j_2}^j) +$$

$$\frac{1}{4}FP(2-P) \sum_{i_1=0}^{T} \sum_{j_1=0}^{T-i_1} \sum_{i_2=0}^{T} \sum_{j_2=0}^{T-i_2} q'(1,i_1,j_1)q'(1,i_2,j_2)(\delta_{i_1}^i + \delta_{i_2}^i)(\delta_{j_1}^j + \delta_{j_2}^j)]/D$$

where $\delta_a^b = 1$ if a = b and $\delta_a^b = 0$ otherwise, and

(4) $D = (1-Px)(1-x) + F[(1-P)(1-Px)x + 2Px(1-x) + Px^2(2-P)]$

Alternatively, if the modifier is on the first chromosome,

$$q''(0,i,j) = [(1-Px)q'(0,i,j) +$$

$$\frac{1}{4}FP \sum_{i_1=0}^{T} \sum_{j_1=0}^{T-i_1} \sum_{i_2=0}^{T} \sum_{j_2=0}^{T-i_2} q'(0,i_1,j_1)q'(1,i_2,j_2)2\delta_{i_1}^i(\delta_{j_1}^j + \delta_{j_2}^j)]/D$$

(3b)

$$q''(1,i,j) = [(1-Px)(1-P)q'(1,i,j) +$$

$$\frac{1}{4}FP \sum_{i_1=0}^{T} \sum_{j_1=0}^{T-i_1} \sum_{i_2=0}^{T} \sum_{j_2=0}^{T-i_2} q'(0,i_1,j_1)q'(1,i_2,j_2)2\delta_{i_2}^i(\delta_{j_1}^j + \delta_{j_2}^j) +$$

$$\frac{1}{4}FP(2-P) \sum_{i_1=0}^{T} \sum_{j_1=0}^{T-i_1} \sum_{i_2=0}^{T} \sum_{j_2=0}^{T-i_2} q'(1,i_1,j_1)q'(1,i_2,j_2)(\delta_{i_1}^i + \delta_{i_2}^i)(\delta_{j_1}^j + \delta_{j_2}^j)]/D$$

(Actually, the formulae (3), although easy to understand, are not suitable for computation. Thus, my program uses the equivalent formulae instead of them).

Finally, the distribution after mutation, $Q(k,i,j)$, which opens the next generation, can be obtained from $q''(k,i,j)$ using:

$$(5)\ Q(k,i,j) = \sum_{i_1=0}^{i} \sum_{j_1=0}^{j} q''(k,i_1,j_1) M(i-i_1) M(j-j_1); \text{ where } M(a) = e^{-U} U^a / a!$$

I have investigated these equations numerically. Apparently, the final frequency of the allele S does not depend on the initial state of the population. An equilibrium frequency of the allele S always increases (or at least does not decrease, if it is exactly 0 or 1) with U. If selection causes the decrease of the frequency of S, it can be accompanied by the increase in the mutation load. Table 1 presents the results on when the three forms of amphimixis invade the apomictic population or eliminate apomixis completely. I have varied U in increments of 0.1 and report its lowest values where invasion or fixation S occurs.

We can see that the spread of amphimixis does not require crossing-over and that all three forms of amphimixis have roughly similar properties. If reduction is not costly, limited recombination actually facilitates the invasion and fixation of amphimixis. Perhaps this is because individual selection favors amphimixis only under substantial mutation load (Kondrashov, 1985), and free recombination, being the most efficient in increasing the variance and reducing the load, thus undermines the conditions that would give amphimixis an advantage. Invasion and fixation of facultative amphimixis (low P) requires much less stringent conditions that those for obligate amphimixis. In most cases there is a substantial range of values of U for which the equilibrium population is polymorphic at the locus S. Thus, the form of amphimixis that is established can be facultative both because of invasion of alleles with incomplete penetrance and because even these alleles may not reach fixation.

If reduction is costly, however, amphimixis with free recombination may invade the apomictic population and become fixed over the widest range of conditions, because when it is advantageous, its advantage is the largest. Still, all forms of amphimixis can invade even when the cost of reduction is high. As in the case of facultative apomixis (Kondrashov, 1985), the cost affects conditions required for fixation of amphimixis much more strongly (and in some cases makes them unrealistic, in terms of the mutation load) than conditions under which the allele S invades the apomictic population but cannot be fixed. Thus, costly reduction probably prevents the evolution of obligate amphimixis.

TABLE 1a - Modifier of amphimixis (two chromosomes with no crossing-over, modifier on one of them) in an apomictic population

	Penetrance							
	0.1		0.4		0.7		1.0	
	Invasion	Fixation	Invasion	Fixation	Invasion	Fixation	Invasion	Fixation
Probability of successful reduction								
1.0	0.1	0.1	0.1	0.1	0.1	0.1	0.1	0.1
0.5	0.2	1.9	0.6	6.4	1.0	9.4	1.5	11.2
0.25	0.3	3.0	0.9	10.0	1.7	14.1	3.0	16.0
0.125	0.3	3.9	1.2	12.9	2.3	17.7	4.4	19.5
0.0625	0.4	4.6	1.3	15.5	2.7	20.7	5.9	22.5

TABLE 1b - Modifier of amphimixis (two chromosomes with no crossing-over, modifier unlinked to both) in an apomictic population

	Penetrance							
	0.1		0.4		0.7		1.0	
	Invasion	Fixation	Invasion	Fixation	Invasion	Fixation	Invasion	Fixation
Probability of successful reduction								
1.0	0.1	0.1	0.2	0.4	0.3	0.6	0.5	0.8
0.5	0.2	1.3	0.6	4.2	1.0	5.9	1.6	6.9
0.25	0.3	2.3	0.8	7.5	1.6	10.3	2.9	11.5
0.125	0.3	3.2	1.0	10.4	2.0	14.0	4.1	15.4
0.0625	0.4	4.0	1.2	13.1	2.5	17.5	5.4	19.0

TABLE 1c - Modifier of amphimixis with free recombination in an apomictic population

	Penetrance							
	0.1		0.4		0.7		1.0	
	Invasion	Fixation	Invasion	Fixation	Invasion	Fixation	Invasion	Fixation
Probability of successful reduction								
1.0	0.1	0.3	0.2	0.8	0.3	1.3	0.5	1.7
0.5	0.2	1.0	0.5	3.4	0.9	4.9	1.4	5.8
0.25	0.2	1.7	0.7	5.4	1.3	7.5	2.4	8.9
0.125	0.3	2.3	0.9	7.3	1.7	10.1	3.4	11.4
0.0625	0.3	2.8	1.0	9.3	2.1	12.7	4.5	14.2

3.3. Regularization of reduction. If the primitive reduction was indeed irregular, its regularization simply abolishes the cost of aneuploidy without any other changes in gene transmission. This can lead to a large immediate advantage, which makes its analysis at the population level unnecessary. Following regularization, amphimixis would then possibly become fixed.

3.4. Origin of crossing-over. Appearance of crossing-over can lead to a further increase of the variance of the per genome number of mutations, and free recombination leads to the lowest mutation load (Fig. 2). However, as in the case of allele S, the invasion and fixation of crossing-over must be considered at the level of individual selection. I assume that crossing-over evolved after reduction became regular, so that cost of aneuploidy is irrelevant here.

Suppose that the allele S with penetrance 1 is already fixed, so that reproduction is obligately amphimictic. The presence of crossing-over depends on a modifier locus C with two alleles C and c. Zygotes cc never have crossing-over, while zygotes Cc and CC have it, which leads to free recombination, with probabilities P_1 and P_2, respectively. The equations which describe the case n = 2 can be easily obtained from the case n = 1 (Kondrashov, 1984).

The data presented in Table 2 show the minimal values of U for which the allele for crossing-over can invade the population or eliminate the allele c completely, either when the locus C is linked to one of two chromosomes or when it is unlinked to both of them (the difference affects only the properties of zygotes where recombination is not free). I call P_2 penetrance and assume that $P_1 = P_2/2$.

TABLE 2a - Modifier of free recombination (on one of two chromosomes) in an amphimictic population with no crossing-over

Penetrance							
0.1		0.4		0.7		1.0	
Invasion	Fixation	Invasion	Fixation	Invasion	Fixation	Invasion	Fixation
0.1	0.1	0.3	0.6	0.4	1.0	0.6	1.3

TABLE 2b - Modifier of free recombination (unlinked to both chromosomes) in an amphimictic population with no crossing-over

Penetrance							
0.1		0.4		0.7		1.0	
Invasion	Fixation	Invasion	Fixation	Invasion	Fixation	Invasion	Fixation
0.8	1.3	0.8	1.5	0.8	1.6	0.8	1.7

The results are similar to the case of the locus S in the amphimixis determining locus presented above. Within some range of U values the allele C invades the population, but is not fixed. Perhaps in this case an allele which causes some intermediate level of crossing-over can be established. Under higher values of U, when the mutation load is about 0.3 or larger, the allele for free recombination can become fixed. Similar results have been obtained previously (Kondrashov, 1984; Charlesworth, 1990), for modifiers of recombination regardless the origin of amphimixis.

4. Discussion

I have considered a possible scenario for the origin of amphimixis from apomixis in three steps: 1) origin of the ploidy cycle consisting of syngamy and primitive reduction, 2) regularization of reduction (if the primitive one was irregular) and 3) evolution of crossing-over. Several cytological mechanisms are possible for every step. Of course, models at the population level do not allow us to determine which mechanisms were actually utilized, because rather different cytology can lead to the same results in terms of transmission genetics. For example, regularization of reduction, regardless of how it occurs, always just abolishes the cost of aneuploidy. This limits the power of our analysis, which cannot shed much light on the origin of the exact cytology of modern amphimixis. However, this is also a strength of such an analysis, because if we limit consideration to the evolution of the mode of inheritance, only a small number of options are possible.

One of the problems that cannot be addressed at the population level is why in modern organisms reduction occurs mostly by two-step meiosis. This process is paradoxical: reduction starts with DNA replication (Hurst, 1993). Apparently, the presence of four DNA molecules is not a priori necessary for crossing-over. One reason why two-step meiosis (or even more bizzare modes of reduction) may be advantageous was recently proposed by Haig (1993). However, from the population point of view, two-step meiosis is simply one of many concievable mechanisms of regular reduction with high-frequency recombination, which defines completely its properties in terms of gene transmission.

The main result of my investigation is that selection against slightly deleterious mutations can favor all the proposed steps in the origin of amhimixis, even if the primitive reduction was costly. Although only strict truncation selection was considered above, the results probably will be similar under any form of synergistic epistasis. The possibility of selection for amphimixis when crossing-over is still absent is particularly important, because the simultaneous origin of the amphimictic ploidy cycle and high-frequency crossing-over seems impossible. Such selection can occur if the ancestral apomict have more than one chromosome in the haploid set.

Establishment of amphimixis as the result of selection against mutations is possible, especially if the primitive reduction was irregular, only under substantial values (of the order of one) of the genomic deleterious mutation rate U and the mutation load L. In unicellular organisms the value of U per cell division is only about 0.003 (Drake, 1991). A population of unicellular organisms cannot withstand L > 0.5 after every cell division, because each individual produces only two offspring. Thus, the conditions required for the origin of amphimixis may be satisfied only if the intensity of selection against mutations fluctuates from generation to generation, so that selection becomes strong only occasionally, when conditions deteriorate. If for a long period of benign conditions the intensity of selection against mutations is low, we can neglect this selection, and consider this period as one "supergeneration". Then, when the conditions deteriorate, the effective U and L can be very high. With present mutation rates, the effective U becomes sufficiently large if conditions deteriorate only every 100 or 1000 generations.

After one generation of amphimixis, the offspring which recieved less than the average number of mutations have their number decreased by approximately $\sqrt{M}$ on the average, where M is the mean per genome number of mutations before amphimixis (with truncation selection M is close to T). After this, during $\sqrt{M}/U$ generations (or supergenerations), the fitness of the apomictic lineages which were started by these offspring remains high. New rounds of amphimixis during this time are not very beneficial, and may even lead to the decline of fitness because of mating with individuals carrying many mutations, while the cost of reduction, if it exists, remains the same. Therefore, obligate amphimixis with free recombination is favored only when $\sqrt{M}/U < 1$ (this condition is roughly equivalent to the genome degradation rate above one, see Kondrashov, 1988) even when there is no cost of reduction. This requires values of U and L too high for one generation of a unicellular organism. If this cost is present, the conditions for establishing of obligate amphimixis are even more stringent. However, facultative amphimixis can be established even with high cost, because rare events of amphimixis may confer a large advantage which lasts many generations.

Therefore, obligate amphimixis cannot evolve in unicellular organisms, at least as the result of selection against mutations. However, amphimixis can be established as a facultative mode of reproduction, if the intensity of selection fluctuates, in which case amphimixis leads to the maximal advantage if it occurs before selection gets stronger. This pattern agrees well with what is observed in nature, where all amphimictic unicellular organisms can also reproduce by mitosis, and amphimixis is often triggered by deterioration or change of conditions (see Margulis *et al.*, 1990).

The cost of aneuploidy would probably have to be overcome during the origin of amphimixis if the primitive reduction had to evolve simultaneously with syngamy, because in this case this reduction can hardly be regular. This would have been the case if the ancestral apomict did not have an endomitotic ploidy cycle. This cost, however, is not prohibitive if n < 5-7 (Fig. 1). Some values of the cost considered above (Table 1) correspond to n > 2, where the advantage of amphimixis should be even higher than with n = 2, although only the case n = 2 was actually studied.

The hypothesis that the ancestral apomict had more than one chromosome is consistent with the data. All contemporary protists studied appear to have n > 1 (Raikov, 1982, Table 3), including those that may be primitively apomictic and whoes ancestors diverged very early (see below). *Giardia lamblia* has 5 chromosomes in the haploid set (see Fan *et al.*, 1991), while *Cyanidioschyzon merolae*, an alga of uncertain affinity with a very small genome (11.7×10^6 nucleotides) has 15 chromosomes (Maleszka, 1993). The cells of various *Euglenida* (see Chaudhary and Prasad, 1986; Prasad and Chaudhary, 1987) and related *Kinetoplastida* (*Trypanosoma brucei*, Van der Ploeg *et al.*, 1989) may have dozens of chromosomes, but diploidy or polyploidy is not excluded (Rawson, 1975).

A cell of *Giardia lamblia* carries perhaps 6 to 10 sets of homologous chromosomes (Fan et al., 1991) and, despite probably being obligately apomictic, somehow maintains very close similarity among them. This suggests that *Giardia lamblia* has an endomitotic ploidy cycle or some other homogenizing mechanism. Such a cycle may be also present in *Amoeba proteus* (Afon'kin, 1986), in *Phaeodaria, Spumellarida, Pyrsonymphida* and some other protists (Raikov, 1982, Ch. 3 and 6; Raikov, 1989, Table 1; Margulis *et al.*, 1990), as well as in some *Phaeophyta* (Muller, 1967) and *Rhodophyta* (see Haig, 1993). In amphimicts this cycle is superimposed on their basic alternation of haploid and diploid phases. If present in the ancestral apomict, an endomitotic ploidy cycle could simplify the origin of amphimixis, because in this case amphimictic reduction could be regular from the very beginning, making the initial cost of aneuploidy and step 2) irrelevant.

An endomitotic ploidy cycle with regular reduction may be present in apomicts simply because the environment sometimes favors haploid, and sometimes - diploid or polyploid cells (Margulis and Sagan, 1986; Hurst and Nurse, 1991). Alternatively, it may evolve as a mutation load-reducing mechanism if selection always favors diploidy or polyploidy. If before selection this cycle always leads to random survival of only one from N genomes of an N-ploid and to restoration of N-ploidy by replicating this genome, the mutation load is $1-e^{-U}$, instead of $1-e^{-NU}$, the value for N-ploid apomicts without ploidy cycle. Even rare, brief returns to haploidy can reduce the load dramatically (Kondrashov, submitted).

If reduction is costly, the alternating advantages of different levels of ploidy caused by a changing environment can hardly cause its evolution (as was suggested by Maguire, 1992), because a better solution seems to be available. Instead of changing the ploidy of a single nucleus and paying the price for inability to do it precisely, the cell could change the number of nuclei, because the properties of a cell with, say, one dipolid nucleus and that with two haploid nuclei are probably very similar. There are numerous examples of cyclical changes of the number of nuclei per cell in many taxa (*Granuloreticulosa, Karyoblastea, Opalinata*, etc., see Margulis *et al.*, 1990; and probably some *Rhyzopoda*, Seravin and Gudkov, 1985; Willumsen *et al.*, 1987). Multinuclear cells can conjugate and exchange nuclei, which may increase the genetic variability to some extent even without true amphimixis, i. e. when the genetic exchange between individual nuclei is impossible (Seravin and Gudkov, 1983, 1984). Only under amphimixis may even irregular reduction be advantageous, because it can lead to a larger increase in the genetic variability than can changes in the number of nuclei or their exchange.

Two other reasons for the origin of the primitive amphimictic ploidy cycle have been proposed. Goodenough (1985) claims that reduction is simply necessary to counterbalance the increase of the ploidy caused by syngamy. This leaves unanswered the question "why did syngamy become a regular event?". Many unicellular forms do not undergo syngamy at any measurable rate. According to another view (see Bernstein and Bernstein, 1991, pp. 295-296) haploidy is generally more efficient, but sometimes diploidy should be briefly maintained, following fusion of two different cells, to repair the DNA damage that cannot be repaired in the absence of an intact copy of the damaged region. However, many unicellular forms are diploid or polyploid (or polynuclear, see above), at least most of the time (e. g. *Bacillariophyta*, *Diplomonadida*, *Actinopoda*, *Ciliata*, see Margulis *et al.*, 1990), so that haploidy is not always the most efficient condition. Thus, the need for the back-up DNA may explain diploidy or polyploidy, but not the ploidy cycle.

The models considered above have assumed that in primitive amphimicts selection occured in the haplophase. This may be the case if the ancestral apomict did not have an endomitotic ploidy cycle. Otherwise, this apomict, as well as its amphimictic derivate, probably were at least partially selected at the diploid or polyploid phase. Still, I think that the effects of recombination in this case would be similar to those under haploid selection. However, under diploid or polyploid selection, in contrast to haploid selection, amphimixis can reduce the mutation load even in the absence of any recombination, i. e. with n = 1 and no crossing-over (Kimura and Maruyama, 1966; Dickson and Manning, 1984; Charlesworth, 1990). This advantage, which also requires synergistic epistasis, appears because segregation can, similar to recombination, create variability in the per genome number of mutations. However, with a high ploidy (or a number of nuclei per cell) the endomitotic ploidy cycle (or occasional reduction of the number of nucei per cell to one) decreases the load more than amphimixis without recombination (or exchange of nuclei between cells) (Kondrashov, submitted). Thus, I do not think that the segregational advantage alone can make the first step of the origin of amphimixis possible, and, therefore, I have assumed that the ancestral apomict had n > 1. However, this is worth further study.

The genetic changes that were necessary for each step in the origin of amphimixis are unknown. However, there is no reason to think that they could not have been achieved by gradual accumulation of new alleles. Origin of primitive syngamy may require only few genetic changes (Seravin and Gudkov, 1984), although egg-sperm interactions in modern organisms can be quite complicated. Origin of reduction and its regularization may require more changes (Maguire, 1992), and high-frequency crossing-over probably required the largest number of them (see Golubovskaya *et al.*, 1993).

Direct paleontological evidence of the modes of reproduction in the ancient eukaryotes is absent. Data on contemporary organisms can elucidate the origin of amphimixis, provided that we know their phylogeny. This phylogeny, although still far from being completely understood, recently became much clearer. Apparently, a majority of contemporary taxa form a "crown" of the eukaryotic tree that have diverged from a common ancestor relatively recently (although still over 1 billion years ago) and almost simultaneously (see Knoll, 1992; Perasso and Baroin-Tourancheau, 1992; Vohra *et al.*, 1992; Martin *et al.*, 1992). These taxa can be grouped into seven "kingdoms": *Chlorophyta* (various green algae, land plants, *Cryptophyta*, and *Acanthamoebidae*; Chapman and Buchheim, 1991; Eschbach *et al.*, 1991); *Chromophyta* (*Bacillariophyta*, *Xanthophyta*, *Phaeophyta*, *Chrysophyta*, *Eustigmatophyta*, *Oomycota*; Bhattacharya *et al.*, 1992); *Rhodophyta* (Bird *et al.*, 1992); *Prymnesiophyta* (Bhattacharya *et al.*, 1993); *Fungi* (*Chytridiomicota*, *Phycomycota*, *Ascomycota*, and *Basidiomycota*; Govind *et al.*, 1992; Bowman *et al.*, 1992; Bruns *et al.*, 1992); *Ciliophora+Apicomplexa+Dynoflagellata* (consisting of these three phyla; the unique features of *Dynophlagellata* seem to be of secondary origin; Gajadhar *et al.*, 1991; Barta *et al.*, 1991; Sadler *et al.*, 1992; Wilson *et al*, 1992), and *Animalia*.

Then, going back into the past, we encounter the divergences (the exact succession is uncertain) of slime molds (cellular - *Dictyostelium*, and plasmodial - *Physarum*; Eschbach *et al.*, 1991); *Kinetoplasitida* and *Euglenida* (which form a single branch); *Archamoebae* (e. g., *Entamoeba histolytica*; Hinkle and Sogin, 1993; Hasegawa *et al.*, 1993); flagellates *Diplomonadida* (e. g., *Giardia lamblia*) and, perhaps, *Retortamonadida*, *Parabasalia*, and *Pyrsonymphida* (Brugerolle, 1993; Cavalier-Smith, 1993; Siddall *et al.*, 1992); and *Microspora* (Schlegel, 1991; Knoll, 1992, Perasso and Baroin-Tourancheau, 1992). The absence of mitochondria, peroxisomes, hydrogenosomes, and Golgi dictyosomes in some of these forms may be primitive (Cavalier-Smith, 1993; Brugerolle, 1993), or may be a secondary trait caused by parasitism (Seravin, 1992).

Thus, the first major question is: "did the common ancestor of the Big Eukaryotic Radiation have amphimixis?". If not, amphimixis is polyphyletic, while otherwise it can be monophyletic at least within the crown, which includes almost all well-studied forms. Currently all seven kingdoms of the crown are overwhelmingly amphimictic. However, it is not clear whether amphimixis was present in each of them from the very beginning. Even apparently the most primitive taxa have amphimixis in *Prymnesiophyta*, *Chromophyta* (*Bacillariophyta*), *Rhodophyta* (*Bangiales*), *Fingi* (*Chytridiomycota*), and *Animalia*. Amphimixis is also obligate in *Ciliata* and *Apicomplexa*, and common in *Dinoflagellata*. It was also reported in some early works on *Prasinophyceae*, perhaps the ancestral group of *Chlorophyta* (see Norris, 1980, pp. 129-130). Thus, we may tentatively conclude that amphimixis in the crown is monophyletic. If so, we must look for its origin even deeper.

Unfortunately, not much is known about the modes of reproduction in the taxa that diverged below the crown. *Dictyostelium* and *Physarum* are amphimictic (Margulis *et al.*, pp. 93, 469), and actually *Dictyostelium* may belong to the crown, rather than split earlier (Loomis and Smith, 1990; Hasegawa *et al.*, 1993). There are few reports of amphimixis in *Euglenida*, while in *Kinetoplastida* it was never seen directly, but genetic evidence suggest that it may exist (see Margulis *et al.*, 1990, pp. 270, 278, 231). Amphimixis occurs in an ameba *Arcella vulgaris*, (*Rhyzopoda*; Mignot and Raikov, 1992), but this group is probably artificial, and other "amebas" can be apomictic. Amphimixis has never been observed in *Diplomonadida* and *Retortamonadida* (Margulis *et al.*, 1990, pp. 204, 246). It was observed in *Pyrsonymphida* (*Oxymonadida*) and *Parabasalia* (*Hypermastogotes*). In the first case meiosis is always one-step, while in the second case both one-step and two-step meiosis was reported (see Raikov, 1982, pp. 202-204). Amphimixis is known in *Microspora*, and meiosis is very unusual (Hazard and Brookbank, 1984; Larsson, 1986, p. 337; Margulis *et al.*, 1990, pp. 57, 59). Thus, even if amphimixis is monophyletic within the crown, it may have originated independently in the more ancient taxa. However, monophyly of amphimixis, i. e. that the common ancestor of all contemporary amphimicts was amphimictic, also cannot be ruled out.

Obviously, we need more data on reproduction in these taxa. Besides, there are many "flagellates", "amebas", "slime molds", etc. whose phylogeny has not been studied using molecular techniques and whose modes of reproduction remain obscure (Patterson and Zolffel, 1991). Some of them may represent other early branches of the eukaryotic tree (Hinkle and Sogin, 1993; Cavalier-Smith, 1993). Many such forms (both apomictic and amphimictic) may be hidden within artificial taxa, and those grouped with unrelated and advanced amphimicts in "amphimictic" taxa can be neglected. Future research will bring a much clearer picture.

If amphimixis is, indeed, polyphyletic, different pathways may have been followed in difference cases, and the overall similarity of meiosis in diffrent taxa may be secondary. Such similarity may appear if in all cases the evolution of meiosis was based on the same ancestral traits. Modern meiosis certainly utilizes, both for segregation and crossing-over, many molecular mechanisms that have first evolved for mitosis, particularly for mitotic DNA repair (Kleckner *et al.*, 1991).

This does not necessarily imply, however, that repair is the function of meiosis, with crossing-over only its unavoidable by-product (see Bernstein and Bernstein, 1991). Recent data contradict this view. Although double-strand DNA breaks are necessary, at least in some cases, to initiate crossing-over, these breaks are not the result of unfortunate accidents, which must be the case if meiosis serves to repair them. Instead, double-strand breaks are deliberately introduced by the cell very shortly before meiosis. Many genes are involved in this process, whose purpose is to initiate chromosome synapsis and crossing-over, and not to repair these breaks, because in the mutants they do not appear in the first place (Sun *et al.*, 1989, 1991; Cao *et al.*, 1990; Padmore *et al.*, 1991; Leem and Ogawa, 1992; Ivanov *et al.*, 1992). Besides, double-strand breaks can be repaired in mitosis without any significant recombination of flanking markers (Nassif and Engels, 1993).

We do not know the exact succession of events in the origin of amphimixis. Perhaps, some cytological details can never be reconstructed. I also do not claim that epistatic selection against mutations is the only possible driving force behind the origin of amphimixis. However, the present analysis shows that the origin of amphimixis "... as the result of a gradual accumulation of changes each one of which had a value as an adaptation ..." (Darlington, 1939, p. 125) is feasible.

References

1. AFON'KIN, S. Ju., Spontaneous 'depolyploidization' of cells in Amoeba clones with increased nuclear DNA content, Archiv fur Protistenkunde **131** (1986), 101-112.
2. BARTA, J. R., M. C. JENKINS and H. D. DANFORTH, Evolutionary relationships of avian *Eimeria* species among other apicomplexan *Protozoa*: monophyly of the apicomplexans supported, Mol. Biol. Evol. **8** (1991), 345-355.
3. BELL, G., 1993. The comparative biology of the alternation of generations, this volume.
4. BENGTSSON, B. O., Deleterious mutations and the origin of the meiotic ploidy cycle, Genetics **131** (1992), 741-744.

5. BERNSTEIN, C. and H. BERNSTEIN, *Aging, Sex, and DNA Repair*, Academic Press, San Diego, (1991).

6. BHATTACHARYA, D., L. MEDLIN, P. O. WAINRIGHT, E. V. ARIZTIA, C. BIBEAN, S. K. STICKEL and L. M. SOGIN, Algae containing chlorophylls a+c are paraphyletic: molecular evolutionary analysis of the *Chromophyta*, Evolution **46** (1992), 1801-1817.

7. BHATTACHARYA, D., S. K. STICKEL and M. L. SOGIN, Isolation and molecular phylogenetic analysis of actin-coding regions from *Emiliania huxleyi*, a prymnesiophyte alga, by reverse transcriptase and PCR methods, Mol. Biol. Evol. **10** (1993), 689-703.

8. BIRD, C. J., E. L. RICE, C. A. MURPHY and M. A. RAGAN, Phylogenetic relationships in the *Gracilariales* (*Rhodophyta*) as determined by 18S rDNA sequences, Phycologia **31** (1992), 510-522.

9. BOWMAN, B. H., J. W. TAYLOR, A. G. BROWNLEE, J. LEE, S.-D. LU and T. J. WHITE, Molecular evolution of the fungi: relationship of the *Basidiomycetes*, *Ascomycetes*, and *Chytridiomycetes*, Mol. Biol. Evol. **9** (1992), 285-296.

10. BRUNS, T. D., R. VILGALYS, S. M. BARNS, D. GONZALEZ, D. S. HIBBETT, D. J. LANE, L. SIMON, S. STICKEL, T. M. SZARO, W. G. WEISBURG and M. L. SOGIN, Evolutionary relationships within the *Fungi*: analyses of nuclear small subunit rRNA sequences, Molecular Phylogenetics and Evolution **1** (1992), 231-241.

11. CAO, L., E. ALANI and N. KLECKNER, A pathway for generation and processing of double-strand breaks during meiotic recombination is *S. cerevisiae*, Cell **61** (1990), 089-1101.

12. CAVALIER-SMITH, T., Evolution and diversity of zooflagellates, J. Eukaryotic Microbiology **40** (1993), 603-605.

13. CHAPMAN, R. L. and M. A. BUCHHEIM, Ribosomal RNA gene sequences: analysis and significance in the phylogeny and taxonomy of green algae, CRC Critical Reviews in Plant Sciences **10** (1991), 343-368.

14. CHARLESWORTH, B., Mutation-selection balance and the evolutionary advantage of sex and recombination, Genet. Res. **55** (1990), 199-221.

15. CHAUDHARY, B. R. and R. N. PRASAD, Contributions to the karyology of euglenoid flagellates. IV. *Thrachelomonas* Ehrenberg emend. Deflandre, Cytologia **51** (1986), 723-729.

16. CLEVELAND, L. R., The origin and evolution of meiosis, Science **105** (1947), 287-289.

17. CROW, J. F., Genetic loads and the cost of natural selection, in *Mathematical Topics in Population Genetics*, edited by K. Kojima. Springer, Heidelberg, (1970), 128-177.

18. DARLINGTON, C. D., *The Evolution of Genetic Systems*, Cambridge University Press, Cambridge, (1939).

19. DICKSON, D. P. E. and J. T. MANNING, Mutational load and the advantage of sex, Heredity **53** (1984), 717-720.

20. DRAKE, J. W., A constant rate of spontaneous mutation in DNA-based microbes, Proc. Natl. Acad. Sci. USA, **88** (1991), 7160-7164.

21. ESCHBACH, S., J. WOLTERS and P. SITTE, Primary and secondary structure of the nuclear small subunit ribosomal RNA of the cryptomonad *Pyrenomonas salina* as inferred from the gene sequence: evolutionary implications, J. Mol. Evol. **32** (1991), 247-252.

22. FAN, J.-B., S. H. KORMAN, C. R. CANTOR and C. L. SMITH, *Giardia lamblia*: haploid genome size determined by pulsed field gel electrophoresis is less than 12 Mb, Nucleic Acids Research **19** (1991), 1905-1908.

23. GAJADHAR, A. A., W. C. MARQUARDT, R. HALL, J. GUNDERSON, E. V. ARIZTIA-CARMONA and M. L. SOGIN, Ribosomal RNA sequences of *Sarcocystis muris*, *Theileria annulata* and *Crypthecodinium cohnii* reveal evolutionary relationships among apicomplexans, dinoflagellates, and ciliates, Molecular and Biochemical Parasitology **45** (1991), 147-154.

24. GOLUBOVSKAYA, I. N., Z. K. GREBENNIKOVA, N. A. AVALKINA and W. F. SHERIDAN, The role of the ameiotic1 gene in the initiation of meiosis and in subsequent meiotic events in maize, Genetics **135** (1993), 1151-1166.

25. GOODENOUGH, U. W., An essay on the origins and evolution of eukaryotic sex, in *The Origin and Evolution of Sex*, edited by H. O. Halvorson and A. Monroy. Alan R. Liss, New York, (1985), 123-140.

26. GOVIND, N. S., K. L. McNALLY and R. K. TRENCH, Isolation and sequence analysis of the small subunit ribosomal RNA gene from the euhaline yeast *Debaryomyces hansenii*, Current Genetics **22** (1992), 191-195.

27. HAIG, D., Alternatives to meiosis: the unusual genetics of red algae, *Microsporidia*, and others, J. of Theor. Biol. **163** (1993), 15-31.

28. HASEGAWA, M., T. HASHIMOTO, J. ADACHI, N. IWABE and T. MIYATA, Early branchings in the evolution of eukaryotes: ancient divergence of *Entamoeba* that lacks mitochondria revealed by protein sequence data, J. Mol. Evol. **36** (1993), 380-388.

29. HAZARD, E. I. and J. W. BROOKBANK, Karyogamy and meiosis in an *Amblyospora sp.* (*Microspora*) in the mosquito *Culex salinarius*. J. of Invertebrate Pathology **44** (1984), 3-11.

30. HEYWOOD, P. and P. T. MAGEE, Meiosis in protists: some structural and physiological aspects of meiosis in algae, fungi, and protozoa, Bacteriological Reviews **40** (1976), 190-240.

31. HINKLE, G. and M. L. SOGIN, The evolution of the *Vahlkampfiidae* as deduced from 16S-like ribosomal RNA analysis, J. Eukaryotic Microbiology **40** (1993), 599-603.

32. HURST, L. D., Evolutionary genetics: drunken walk of the diploid, Nature **365** (1993), 206-207.

33. HURST, L. D. and P. NURSE, A note on the evolution of meiosis, J. Theor. Biol. **150** (1991), 561-563.

34. IVANOV, A. V., On the independent and convergent origin of sexual process, Monitore Zoologico Italiano n. ser. **4** (1970), 183-189.

35. IVANOV, E. L, V. E. KOROLEV and F. FABRE, XRS2, a DNA repair gene of *Saccharomyces cerevisiae*, is needed for meiotic recombination, Genetics **132** (1992), 651-664.

36. JOHN, B., *Meiosis*. Cambridge University Press, Cambridge, (1990).

37. KIMURA, M. and T. MARUYAMA, The mutational load with epistatic gene interactions in fitness, Genetics **54** (1966), 1337-1351.

38. KLECKNER, N., R. PADMORE and D. K. BISHOP, Meiotic chromosome metabolism: one view, Cold Spring Harbor Symp. Quant. Biol. **56** (1991), 729-743.

39 KNOLL, A. H., The early evolution of eukaryotes: a geological perspective, Science **256** (1992), 622-627.

40. KONDRASHOV, A. S., Selection against harmful mutations in large sexual and asexual populations, Genet. Res. **40** (1982), 325-332.

41. KONDRASHOV, A. S., Deleterious mutations as an evolutionary factor. I. The advantage of recombination., Genet. Res **44** (1984), 199-217.

42. KONDRASHOV, A. S., Deleterious mutations as an evolutionary factor. II. Facultative apomixis and selfing, Genetics **111** (1985), 635-653.

43. KONDRASHOV, A. S., Deleterious mutations and the evolution of sexual reproduction, Nature **336** (1988), 435-440.

44. KONDRASHOV, A. S., Classification of hypotheses on the advantage of amphimixis, J. of Heredity **84** (1993), 372-387.

45. LARSSON, R., Ultrastructure, function, and classification of *Microsporidia*. in *Progress in Protozoology*, edited by J. O. Corliss and D. J. Patterson. Biopress Ltd., Bristol, **1** (1986), 325-390.

46. LEEM, S.-H. and H. OGAWA, The MRE4 gene encodes a novel protein kinase homologue required for meiotic recombination in *Saccharomyces cerevisiae*, Nucleic Acids Research **20** (1992), 449-457.

47. LOOMIS, W. F. and D. W. SMITH, Molecular phylogeny of *Dictyostelium discoideum* by protein sequence comparison, Proc. Natl Acad. Sci. USA **87** (1990), 9093-9097.

48. MAGUIRE, M. P., The evolution of meiosis, J. Theor. Biol. **154** (1992), 43-55.

49. MALESZKA, R., Electrophoretic analysis of the nuclear and organellar genomes in the ultra-small alga *Cyanidioschyzon merolae*, Current Genetics **24** (1993), 548-550.

50. MARGULIS, L. and D. SAGAN, *Origins of Sex*, Yale University Press, New Haven (1986).

51. MARGULIS, L., J. O. CORLISS, M. MELKONIAN and D. J. CHAPMAN, eds, *Handbook of Protoctista*, Jones and Bartlett, Boston, (1990).

52. MARTIN, W., C. C. SOMERVILLE and S. LOISEAUX-DE GOER, Molecular phylogenies of plastid origins and algal evolution, J. Mol. Evol. **35** (1992), 385-404.

53. McKEE, B. D., L. HABERA and J. A. VRANA, Evidence that intergenic spacer repeats of *Drosophila melanogaster* rRNA genes function as X-Y pairing sites in male meiosis, and a general model for achiasmatic payring, Genetics **132** (1992), 529-544.

54. MIGNOT, J.-P. and I. B. RAIKOV, Evidence for meiosis in the testate amoeba *Arcella*, J. of Protozoology **39** (1992), 287-289.

55. MULLER, D. G., Generationswechel, Kernphasenwechesel and Sexualitat der Braunalge *Ectocarpus siliculosus*, Planta **75** (1967), 39-54.

56. NASSIF, N. and W. ENGELS, DNA homology requirements for mitotic gap repair in *Drosophila*, Proc. Natl. Acad. Sci. USA **90** (1993), 1262-1266.

57. NORRIS, R. E., *Prasinophytes*. in *Phytoflagellates*, edited by E. R. Cox. Elsevier, New York, (1980), 85-145.

58. OAKLEY, H. A, Meiosis in *Mesostoma ehrenbergii ehrenbergii* (*Turbellaria, Rhabdocoela*) III. Univalent chromosome segregation during the first meiotic division in spermatocytes, Chromosoma **91** (1985), 95-100.

59. PADMORE, R., L. CAO and N. KLECKNER, Temporal comparison of recombination and synaptonemal complex formation during meiosis in *S. cerevisiae*, Cell **66** (1991), 1239-1256.

60. PATTERSON, D. J. and M. ZOLFFEL, Heterotrophic flagellates of uncertain taxonomic position. in *The Biology of Free-living Heterotrophic Flagellates*, edited by D. J. Patterson and J. Larson. Clarendon Press, Oxford, (1991), 427-476.

61. PENNY, D., The evolution of meiosis and sexual reproduction, Biol. J. Linnean Soc. **25** (1985), 209-220.

62. PERASSO, R. and A. BAROIN-TOURANCHEAU, L'Eucaryogenese: un modele derive des phylogenies moleculaires basees sur les ARN ribosomiques, Comptes Rendus des Seances Soc. Biol. **186** (1992), 656-665.

63. PRASAD, R. N. and B. R. L. Chaudhary, Contributions to the karyology of euglenoid flagellates. II. *Lepocinclis* Petry, Cytologia **52** (1987), 357-360.

64. RAIKOV, I. B., *The Protozoan Nucleus,* Springer, Berlin, (1982).

65. RAIKOV, I. B., Nuclear genome of the *Protozoa*. in *Progress in Protozoology*, edited by D. J. Patterson and J. O. Corliss. Biopress Ltd., Bristol, **3** (1989), 21-86.

66. RAWSON, J. R. Y., The characterization of *Euglena gracilis* DNA by its reassociation kinetics, BBA **402** (1975), 171-178.

67. SADLER, L. A., K. L. McNALLY, N. S. GOVIND, C. F. BRUNK and R. K. TRENCH, The nucleotide sequence of the small subunit ribosomal RNA gene from *Symbiodinium pilosum*, a symbiotic dinoflagellate, Current Genetics **21** (1992), 409-416.

68. SCHLEGEL, M., Protist evolution and phylogeny as discerned from small subunit ribosomal RNA sequence comparisons, Europ. J. of Protistology **27** (1991), 207-219.

69. SERAVIN, L. N., Eukaryotes lacking the most important cellular organelles (flagella, Golgi complex, mitochondria), and the main purpose of organellology, Tsitologiya **34** (1992), 3-33.

70. SERAVIN, L. N. and A. V. GUDKOV, Spontaneous cell fusion in culture of marine amoeba *Hyperamoeba fallax*, Tsitologiya **25** (1983), 194-199.

71. SERAVIN, L. N. and A. V. GUDKOV, The main types and forms of agamic cell fusion in *Protozoa*, Tsitologiya **26** (1984), 123-131.

72. SERAVIN, L. N. and A. V. GUDKOV, Similarity and difference between two marine limax amoebae, *Gruberella flavescens* and *Euhyperamoeba fallax* (*Lobosea, Gymnamoebia*), Zoologicheskij Zhurnal **64** (1985), 1090-1093.

73. SIDDALL M. E., H. HONG, S. S. DESSER, Phylogenetic analysis of the *Diplomonadida* (Wenyon, 1926) Brugerolle, 1975: evidence for heterochrony in

Protozoa and against *Giardia lamblia* as a "missing link", J. of Protozoology **39** (1992), 361-367.

74. SUN, H., D. TRECO and J. W. SZOSTAK, Double-strand breaks at an initiation site for meiotic gene conversion, Nature **338** (1989), 87-90.

75. SUN, H., D. TRECO, N. P. SCHULTES and J. W. SZOSTAK, Extensive 3'-overhanging, single-stranded DNA associated with the meiosis-specific double-strand breaks at the ARG4 recombination initiation site, Cell **64** (1991), 1155-1161.

76. UYENOYAMA, M. K. and B. O. BEHGTSSON, On the origin of meiotic reproduction: a genetic modifier model, Genetics **123** (1989), 873-885.

77. VAN der PLOEG, L. H. T., C. L. SMITH, R. I. POLVERE and K. M. GOTTESDIENER, Improved separation of chromosome-sized DNA from *Trypanosoma brucei*, stock 427-60, Nucleic Acids Research **17** (1989), 3217-3227.

78. WILLUMSEN, N. B. S., F. SIEMENSMA and P. SUHR-JESSEN, A multinucleate amoeba, *Parachaos zoochlorellae* (Willumsen 1982) *comb. nov.*, and a proposed division of the genus *Chaos* into the genera *Chaos* and *Parachaos* (*Gymnamoebia*, *Amoebidae*), Archiv fur Protistenkunde **134** (1987), 303-313.

79. WILSON, R. J. M., M. FRY, M. J. GARDNER, J. E. FEAGIN and D. H. WILLIAMSON, Subcellular fractionation of the two organelle DNAs of malaria parasites, Current Genetics **21** (1992), 405-408.

SECTION OF ECOLOGY AND SYSTEMATICS
CORNELL UNIVERSITY
ITHACA, NY 14853

alex_kondrashov@qmrelay.mail.cornell.edu

Lectures on Mathematics in the Life Sciences
Volume **25**, 1994

Deleterious Mutation and Ecological Selection in the Evolution of Life Cycles

C. D. JENKINS and M. KIRKPATRICK

ABSTRACT. One classical genetic explanation for the evolution of life cycles is that diploids have an advantage due to their ability to mask the effects of deleterious mutations. This advantage may be transient, however, as diploids also have twice the mutation rate of haploids. This doubled mutation rate leads to reduced population mean fitness at equilibrium for diploids and to questions as to the actual evolutionary advantage to diploidy. We present a model of life cycle evolution where the evolutionary force is selection against deleterious mutations at a single locus, and the relative length of haploid and diploid phases is under genetic control. Even though an increase in the diploid phase leads to a reduction in population mean fitness, we find that diploidy is favored under the biologically reasonable conditions of weak selection, partial dominance of the mutation and positive recombination. Haploidy is favored when there is complete linkage or when both selection and dominance are strong. The strength of selection on the life cycle is very weak in this case, on the order of the per locus mutation rate, meaning that the results may be very sensitive to the presence of other evolutionary forces. We therefore analyze the effect of adding ecological selection to the model. The ecological selection considered is selection acting on differences between the phases in their constant mortality rates. Under this model, ecological selection overwhelms selection against mutations, leading to the evolution of whichever life cycle has the lowest mortality rate. We conclude by summarizing the results of a companion study that considers the effect of selection against mutations across the entire genome. Life cycles evolve as they did under the single locus model, but the strength of selection on the life cycle under this multilocus model is on the order of the per genome mutation rate. Therefore, the effect of selection against deleterious mutations may be substantial when the entire genome is considered.

1991 Mathematics Subject Classification: 92D15

This work was supported by NIH (1-R55-GM45226-01) and NSF (BSR-9107140) grants to
  M. K. We are grateful to A. Kondrashov for helpful discussions.

This paper is in final form and no version of it will be submitted for publication elsewhere.

Introduction

As a direct consequence of sexual reproduction, most organisms pass through both a haploid and a diploid phase during their life cycle. The relative duration of each phase varies greatly across taxa, ranging from haploid-dominated cycles in many protists and fungi to diploid-dominated cycles in animals, with a wide diversity of life cycles in organisms that retain both phases. The evolution of this diversity is still poorly understood (see Valero *et al.,* 1992 and Bell, this volume). Explanations offered generally fall into two broad categories, genetic and ecological. The first of these stresses the genetical consequences of having one *vs.* two homologous genomes. The second category considers ecological and physiological differences between phases as the driving force behind life cycle evolution. In this paper, we will consider an often invoked genetic explanation for an advantage to the diploid phase: selection against deleterious mutations favors diploids because of their ability to protect against the full expression of these mutations. Our objective is to determine how life cycles evolve under this type of selection. We will then consider a simple case that includes both ecological selection and selection against deleterious mutations in order to study the relative effects of these two forces on the evolution of life cycles.

The classical genetical explanation of the evolution of diploidy is that a diploid genome allows for complementation and thus for protection against the expression of deleterious mutations (Muller, 1932; Crow & Kimura, 1965; Bernstein, Byers & Michod, 1981). This masking ability provides an advantage to a diploid individual over a haploid one if they have the same per locus probability of carrying a mutation, but this diploid advantage may be transient. Because diploids have two copies of every gene, they also have twice the total mutation rate of haploids. Thus, as partially recessive mutations accumulate, the mean fitness of a diploid population is reduced until, at equilibrium, a diploid population has approximately twice the mutation load of a haploid population. Diploidy therefore has an advantage due to masking but a cost due to a doubled mutation rate, relative to haploidy. The evolutionary question of interest is to determine when the advantages to diploidy outweigh the disadvantages, leading to diploid life cycles. In addition, it is important to consider whether life cycles that retain both phases ever evolve as a balance between these effects.

Previous theoretical studies have not reached a consensus on the first of these issues. In a model of genome-wide deleterious mutation, Kondrashov & Crow (1991) considered whether an all-haploid or an all-diploid life cycle would have highest mean fitness. They concluded that haploidy is favored if loci are selected independently, but diploidy may give the highest fitness with strong epistasis. In contrast, Perrot, Richerd & Valero (1991), Bengtsson (1992) and Otto & Goldstein (1992) considered a single locus with mutation to a deleterious allele and the evolution of a modifier locus that causes individuals to be selected entirely as haploids or entirely as diploids. In this situation, diploidy is favored under reasonable conditions regarding dominance and recombination. None of these analyses allow for the possibility of mixed life cycles.

The results from these previous studies present a paradox: they make opposite predictions in the absence of strong epistasis. It is unclear whether this discrepancy is due to different assumptions about the biology of the life cycle or the number of loci considered, or to differences between a mean fitness and an individual selection analysis.

Our goal in this paper is to resolve the apparent contradiction. We will first focus on a model of a single locus mutating to a deleterious allele, and will also review results from a companion study of the situation where a very large number of loci throughout the genome mutate to deleterious alleles (Jenkins & Kirkpatrick, unpublished manuscript). In these analyses, a life cycle of fixed length evolves so that different proportions of each generation are spent in haploid and diploid phases. Thus we consider not the evolution of life cycle length but of the relative proportions spent in each phase, making life cycles that retain both phases possible under this model. Since selection acts in each phase of the life cycle, this approach reflects the idea that the length of each genetic phase may be the result of a balance between the advantages of being haploid and of being diploid.

To date, models that consider selection against deleterious mutations as an explanation for the evolution of life cycles have always assumed there is no ecological selection – haploids and diploids are identical except for the potential for masking in diploids. It would be interesting to know how sensitive the results obtained are to this assumption. With this objective in mind, we will analyze a second model that includes a simple form of ecological selection to determine which of the two effects is likely to determine how life cycles evolve.

The Models

Selection against deleterious mutations (Model 1). Consider an infinite population of meiotically reproducing organisms whose genetic life cycle is of fixed duration. Each individual spends a proportion t of its life cycle in the haploid phase, with $(1 - t)$ defining the proportion spent as a diploid (Figure 1). Gametes undergo random mating, followed by diploid selection, recombination, mutation, and haploid selection.

Assume the life cycle is under the genetic control of a single locus M. In this model, generations are non-overlapping and the cycle is synchronized by some external seasonal event. It is reasonable to assume that this synchrony occurs at the time of syngamy in order to assure a high probability of fertilization. In this case, the life cycle genotype in the diploid phase controls the timing of meiosis and the succeeding haploid phase lasts until the next synchronous mating event, regardless of the haploid genotype.

The objective of this study is to determine genetically stable life cycles. Our strategy is to assume the population is initially fixed for allele M which gives some life cycle t_1. We then determine whether allele m can spread in this population when heterozygotes Mm give life cycle t_2.

The single viability locus, which is expressed in both the haploid and diploid phases, has two alleles, A and a, with mutation from A to a at a rate u; back mutation is assumed to be negligible. Viability in each phase is a function of the time spent in that phase, with an individual's mortality rate depending on the genotype at locus A and whether it is in haploid or diploid phase. Mortality rates are constant over time within each phase.

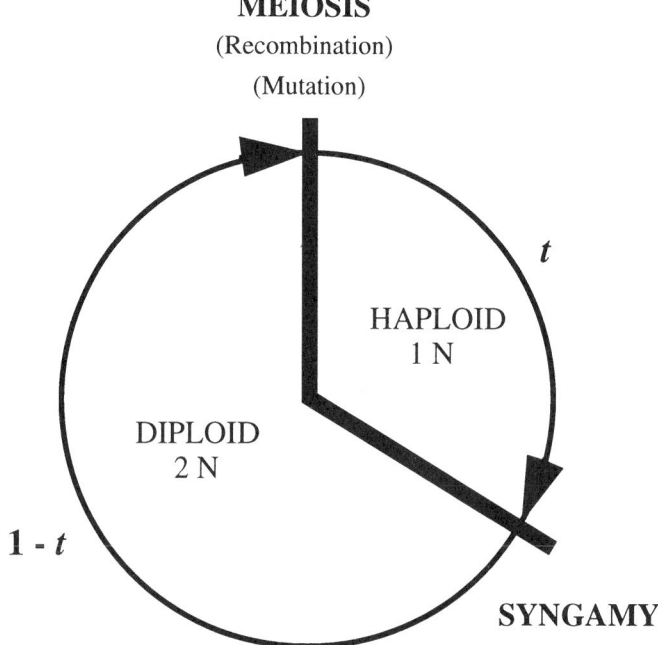

FIGURE 1 - Life cycle. Selection in each phase is a function of t, where t is the proportion of the life cycle spent in the haploid phase.

Relative fitnesses under this model of selection and mutation are given in Table 1a. Haploid and diploid wild-type individuals are assumed to have the same mortality rate. This means that, without loss of generality, we can assign these genotypes (AA and A) a fitness of 1 and indicate the fitnesses of the mutant genotypes by their mortality rate relative to that of the wild-type. Individuals that have only mutant alleles (aa and a) have a mortality rate s, while heterozygotes (Aa) have mortality rate hs, where h is a measure of the dominance of the mutation. We assume both h and s are positive and $0 < h \leq 1$. When h is less than 1/2, the heterozygote has a viability closer to wild-type than to mutant individuals, and masking of deleterious effects occurs in diploids. Note that this fitness scheme assumes there are no differences in mortality rates between haploids and diploids other than those having to do with the genotype at the selected locus; there is no ecological selection in this model. This assumption allows us to determine the evolutionary impact of selection against deleterious mutations in isolation of ecological effects.

Table 1 - Fitnesses

General fitness notation (used in equation 1)							
HAPLOID FITNESS				**DIPLOID FITNESS**			
Parental genotype* MM	Mm	mm			MM	Mm	mm
Haploid genotype	M	M and m	m	AA	w_{11}	w_{21}	w_{31}
A	w_1	w_3	w_5	Aa	w_{12}	w_{22}	w_{32}
a	w_2	w_4	w_6	aa	w_{13}	w_{23}	w_{33}

* Diploid parental genotype controls haploid life cycle phenotype.

Fitnesses:	
a. Model 1 - Selection against deleterious mutations	**b. Model 2 - Mutations and ecological selection**

Life cycle genotype MM

Model 1	Model 2
$w_1 = 1$	$w_1 = \exp[-\mu_h t_1]$
$w_2 = \exp[-st_1]$	$w_2 = \exp[-(\mu_h + s)t_1]$
$w_{11} = 1$	$w_{11} = \exp[-\mu_d(1 - t_1)]$
$w_{12} = \exp[-hs(1 - t_1)]$	$w_{12} = \exp[-(\mu_d + hs)(1 - t_1)]$
$w_{13} = \exp[-s(1 - t_1)]$	$w_{13} = \exp[-(\mu_d + s)(1 - t_1)]$

Life cycle genotype Mm

Model 1	Model 2
$w_3 = 1$	$w_3 = \exp[-\mu_h t_2]$
$w_4 = \exp[-st_2]$	$w_4 = \exp[-(\mu_h + s)t_2]$
$w_{21} = 1$	$w_{21} = \exp[-\mu_d(1 - t_2)]$
$w_{22} = \exp[-hs(1 - t_2)]$	$w_{22} = \exp[-(\mu_d + hs)(1 - t_2)]$
$w_{23} = \exp[-s(1 - t_2)]$	$w_{23} = \exp[-(\mu_d + s)(1 - t_2)]$

Life cycle genotype mm

Model 1	Model 2
$w_5 = 1$	$w_5 = \exp[-\mu_h t_3]$
$w_6 = \exp[-st_3]$	$w_6 = \exp[-(\mu_h + s)t_3]$
$w_{31} = 1$	$w_{31} = \exp[-\mu_d(1 - t_3)]$
$w_{32} = \exp[-hs(1 - t_3)]$	$w_{32} = \exp[-(\mu_d + hs)(1 - t_3)]$
$w_{33} = \exp[-s(1 - t_3)]$	$w_{33} = \exp[-(\mu_d + s)(1 - t_3)]$

$0 \leq s \leq 1$, $0 \leq h \leq 1$, $0 \leq \mu_h \leq 1$, $0 \leq \mu_d \leq 1$, and $0 \leq t_i \leq 1$.

Let x_1, x_2, x_3, and x_4 represent the frequencies of the MA, Ma, mA and ma gametes after haploid selection and r the recombination fraction between the M and A loci. Gamete frequencies the next generation are:

$$x_1' = (w_1(w_{11}x_1^2 + w_{12}x_1x_2) + w_3(w_{21}x_1x_3 + w_{22}x_1x_4 - rw_{22}D))(1-u)/\overline{W}$$

$$x_2' = (\ (w_2(w_{11}x_1^2 + w_{12}x_1x_2) + w_4(w_{21}x_1x_3 + w_{22}x_1x_4 - rw_{22}D))\ u$$
$$+ (w_2(w_{12}x_1x_2 + w_{13}x_2^2) + w_4(w_{22}x_2x_3 + w_{23}x_2x_4 + rw_{22}D))\ )/\overline{W}$$

(1)

$$x_3' = (w_5(w_{31}x_3^2 + w_{32}x_3x_4) + w_3(w_{21}x_1x_3 + w_{22}x_2x_3 + rw_{22}D))(1-u)/\overline{W}$$

$$x_4' = (\ (w_6(w_{31}x_3^2 + w_{32}x_3x_4) + w_4(w_{21}x_1x_3 + w_{22}x_2x_3 + rw_{22}D))\ u$$
$$+ (w_6(w_{32}x_3x_4 + w_{33}x_4^2) + w_4(w_{22}x_1x_4 + w_{23}x_2x_4 - rw_{22}D))\ )/\overline{W}$$

where

$$D = x_1x_4 - x_2x_3 \quad \text{and}$$

$$\overline{W} = \text{sum of the numerators.}$$

We first find the equilibrium values for x_1 and x_2 assuming allele M is fixed in the population ($x_3 = x_4 = 0$). We then consider the dynamics of this system when allele m is introduced at low frequency by performing a linear stability analysis around this equilibrium.

Mutations and ecological selection (Model 2). Consider a model identical to that discussed previously with a single exception: in addition to selection against deleterious mutations, we also allow the wild-type viability in the haploid and diploid phases to differ. Such a difference between the phases could be due to different morphologies, niches or inherent differences in haploid and diploid cells (such as size or growth rates). Fitnesses under this model of selection and mutation are given in Table 1b. Haploid and diploid individuals have different baseline mortality rates, regardless of the genotype at the viability locus, with additional selection against the mutation identical to that in Model 1. It can be shown that any differences between the haploid and diploid phase in fecundity or in non-time dependent viability will not affect the equilibria or the dynamics of this system. The fitnesses from Table 1b are substituted into recursion equations 1 and the strategy is again to determine genetically stable life cycles using the linear stability analysis outlined above.

Results

Model 1 – Selection against deleterious mutations.   When the M allele is fixed $(x_3 = x_4 = 0, x_1 = 1 - x_2)$, the equilibrium frequency of the deleterious allele at the viability locus is

$$(2) \qquad\qquad x_2 \approx \frac{u \exp[-s\,t_1]}{(1 - \exp[-hs - t_1(s - hs)])}$$

for $h > 0$. We have assumed that terms of the order of the square of the mutation rate can be ignored.   In the limiting case where both $h = 0$ and $t = 0$, mutations are fully recessive and $x_2 = \sqrt{u}/(1 - \exp[-s])$.   Since there is strong empirical evidence that most deleterious mutations are partially rather than fully recessive (Simmons & Crow, 1977), we will consider only the case of $0 < h < 1$ for the remainder of this analysis.

At this equilibrium, population mean fitness is:

$$(3) \qquad\qquad \overline{W} \approx 1 - \frac{u\,(1 + \exp[-s t_1] - 2\exp[-hs - t_1(s - hs)])}{(1 - \exp[-hs - t_1(s - hs)])}$$

and is maximized by $t_1 = 1$;   haploidy gives the highest population mean fitness.   As predicted by standard theory (Crow & Kimura, 1970), when s is small, the mutation load of a completely haploid population ($t = 1$) is approximately u.   The load of a completely diploid population ($t = 0$) with partially recessive mutations ($h > 0$) is approximately $2u$.

The stability of this equilibrium to changes in life cycle is characterized by the leading eigenvalue (λ_1) found from the linear stability analysis and given by equation 4 (next page).   When the magnitude of this eigenvalue is less than 1, the original equilibrium with allele M fixed is stable to invasion by a new life cycle allele m;   a magnitude greater than 1 indicates that the new allele increases in frequency.   Since we have assumed that s and h are positive and that $h < 1$, the constant K_2 in equation 4 is positive, so the sign of K_1 determines the stability of the system.   If $K_1 > 0$, $\lambda_1 < 1$ and the initial equilibrium is stable.   If $K_1 < 0$, $\lambda_1 > 1$ and the new life cycle allele spreads.   For arbitrary values of recombination, K_1 is a complicated function of the difference in life cycles as well as the selection and dominance parameters and, in general, requires numerical analysis to obtain results.

In the special case where the life cycle locus and the viability locus are completely linked ($r = 0$), however, we find a simple result.   Any allele that increases the proportion of the life cycle spent as a haploid will spread and only complete haploidy is stable.   This result follows from the fact that, when $r = 0$, the sign of A in equation 4 determines the stability of the system.   We can see that whenever the initial life cycle has a longer haploid phase than the invading life cycle (that is, when $t_1 > t_2$) the initial equilibrium is stable.   If $t_1 < t_2$, the new allele will spread.   This result indicates that when there is complete linkage between the viability and modifier locus, the life cycle that gives maximum population mean fitness is the only stable life cycle.

We will see below that mean fitness is not necessarily maximized when the two loci can recombine.

$$(4) \qquad \lambda_1 \approx 1 - \frac{u\,K_1}{(1 - \exp[-hs - t_1(s - hs)])\,K_2}$$

where:

$$K_1 = A\,(1 - \exp[-hs - t_2(s - hs)])$$

$$+ r(B\,\exp[-hs(1 - t_2)] + 2A\,\exp[-hs - t_2(s - hs)])$$

$$K_2 = 1 - (1 - r)\exp[-hs - t_2(s - hs)]$$

$$A = \exp[-st_1](\exp[-hs(1 - t_1)] - \exp[-hs(1 - t_2)])$$

$$B = \exp[-st_1] - \exp[-st_2]$$

The strength of selection on the life cycle can be quantified by the effective selection coefficient, defined as $s_{m1} = \lambda_1 - 1$. A rare life cycle allele evolves at the same rate as would an allele under simple viability selection that has a selection coefficient s_{m1}. A simple expression for s_{m1} can be found when the change in t caused by the rare allele is small. Denoting this change as $\delta t = t_2 - t_1$, a Taylor series expansion of equation 4 gives the following:

$$(5) \qquad s_{m1} \approx \frac{\delta t\, u\, s\, \exp[-hs - t_1(s - hs)]\,K_3}{(1 - \exp[-hs - t_1(s - hs)])\,K_4}$$

where:

$$K_3 = h - r - h(1 - 2r)\exp[-hs - t_1(s - hs)]$$

$$K_4 = 1 - (1 - r)\exp[-hs - t_1(s - hs)]$$

K_4 is positive, so the sign of s_{m1} is determined by the sign of $\delta t\, K_3$. A positive value of δt indicates that the new allele shifts the life cycle toward a longer haploid phase, while a negative δt means more diploidy. Therefore, if $K_3 > 0$, life cycle alleles that increase the haploid phase will spread while those that increase diploidy will not. The opposite is true if $K_3 < 0$; alleles that increase the diploid phase are favored. Figure 2 illustrates the stability of different life cycles for two values of the recombination rate, r.

Several results from this analysis merit further comment. First, we find that a simple condition describes the stability of the system when there is free recombination between the life cycle and the viability locus ($r = 1/2$; Figure 2a). If there is any degree of masking in the heterozygote, alleles that increase the diploid phase will always spread, regardless of the initial life cycle, and diploid life cycles are stable. In this case, $K_3 > 0$ whenever $h > 1/2$, and $K_3 < 0$ whenever $h < 1/2$.

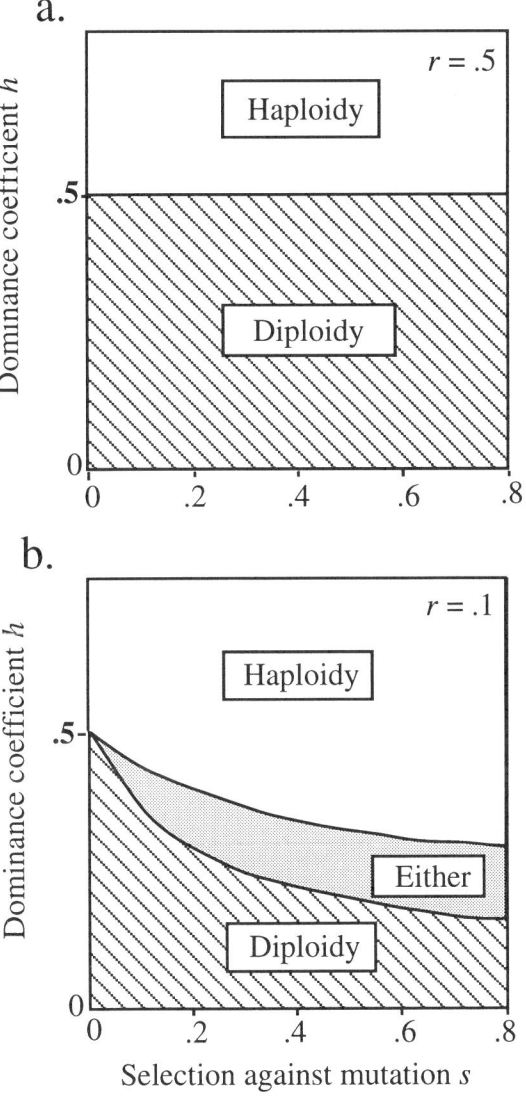

FIGURE 2. Evolution of life cycles under selection against deleterious mutations. The striped area indicates parameters for which alleles that increase diploidy are favored and the white area indicates that increased haploidy is favored. In the dotted region, either an increase in haploidy or in diploidy will be favored, depending on the initial life cycle of the population.   a. Free recombination between the viability and life cycle loci. b. Partial recombination ($r = 0.1$).

Second, for values of recombination between 0 and 1/2, increased diploidy is favored when both selection and dominance are small, and increases in the haploid phase are favored under conditions of strong selection or high dominance. This result is illustrated for $r = 0.1$ in Figure 2b.

Third, for some combinations of selection and dominance, stable diploid and stable haploid life cycles are possible (Figure 2b). For example, when $r = 0.1$, $s = 0.2$, $h = 0.3$ and the initial life cycle $t_1 = 0.875$, new life cycle alleles that increase the haploid phase and those that increase the diploid phase will both spread. In this case, both haploidy and diploidy are stable to invasion by new alleles of small effect. The life cycle that results will depend on the initial frequencies of alleles in the population and on the initial life cycle.

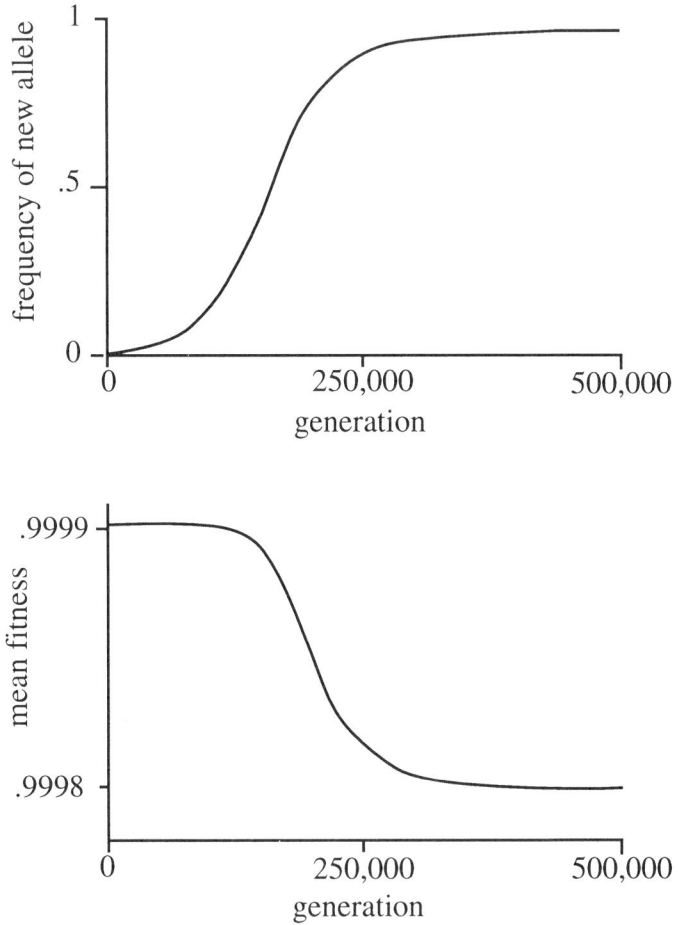

FIGURE 3. Invasion of a rare life cycle allele and the corresponding reduction in population mean fitness. Parameters used: $s = 0.1$, $h = 0.2$, $r = 0.5$, $u = 10^{-4}$, $t_1 = 1$, $t_2 = 0$.

Fourth, this analysis shows that evolution does not always maximize mean fitness. Alleles that increase the diploid phase may spread even though they reduce population mean fitness. An example of this effect is illustrated in Figure 3. In this example, $s = 0.1$, $h = 0.2$, $u = 10^{-4}$ and $r = 0.5$. An allele that increases the diploid phase is introduced into a haploid population. As the new allele spreads, a larger fraction of the population becomes diploid, mutations begin to accumulate at a higher rate and the mean fitness of the population decreases.

Fifth, a numerical analysis of λ_1 and s_{m1} shows that there are some combinations of parameters for which alleles that give a small change in life cycle will not spread, but modifiers of large effect will. For example, if there is free recombination ($r = 1/2$), $s = 0.5$ and $h = 0.49$, a diploid population will be stable to invasion by alleles that give a new life cycle with $t = 0.1$ (a slight increase in the haploid phase), but new alleles that give an entirely haploid life cycle ($t = 1$) will spread. Increased diploidy alleles will be able to spread in an entirely haploid population, however, so haploidy is not stable. This difference between the evolution of modifiers of small and large effect appears to be confined to values of dominance close to those that result in neutral stability.

Sixth, the strength of selection on the new life cycle allele (the magnitude of s_{m1}) is more strongly influenced by u, the per-locus mutation rate, than by s, the strength of selection against the mutation. The value of s_{m1} is on the order of this mutation rate times δt, the size of the life cycle change from the new allele (see equation 5). The sign of s_{m1} is not a function of u, however; the stability of the initial equilibrium is not affected by the rate at which deleterious mutations appear.

Finally, numerical analysis shows that mixed life cycles (those that include both phases ($0 < t < 1$)) are not stable under this model.

Model 2 - Mutations and ecological selection. There are two distinct selective forces at work in this model, differential mortality on the haploid and diploid phases themselves and selection against deleterious mutations. Regardless of the values of the baseline haploid and diploid mortality rates, the equilibrium frequency of the deleterious allele at the viability locus when life cycle allele M is fixed ($x_3 = x_4 = 0$) is again given by the expression in equation 2. Population mean fitness and the dynamics of this model do, however, depend on both selective forces. At the equilibrium described in equation 2, population mean fitness is:

$$(6) \quad \overline{W} \approx \exp[-\mu_d - t_1(\mu_h - \mu_d)]\left(1 - \frac{u(1 + \exp[-st_1] - 2\exp[-hs - t_1(s - hs)])}{(1 - \exp[-hs - t_1(s - hs)])}\right)$$

The first term of equation 6 corresponds to the effect on mean fitness of ecological selection. It increases with the amount of time spent in the phase with lower mortality. The term in brackets is the contribution to mean fitness of selection against the mutation. This term is identical to $\overline{W}$ from Model 1 (equation 3). Whenever $\| \mu_h - \mu_d \| \gg u$, the first term will have the controlling effect on mean fitness. Therefore, whenever there is a substantial difference between the mortality rates of the two phases, mean fitness will be maximized by the life cycle, either all-haploid or all-diploid, that has the lower mortality rate.

The stability of this equilibrium is again determined by the leading eigenvalue found from the linear stability analysis, now given by

$$(7) \qquad \lambda_2 \approx \lambda_1 \ \exp[(\mu_h - \mu_d)(t_1 - t_2)]$$

where λ_1 is the eigenvalue found in Model 1 (equation 4). The expression for λ_2 is complicated and, in general, requires numerical analysis to obtain results. We will focus again on the effective selection coefficient of the new life cycle allele when the allele has a small effect:

$$(8) \qquad s_{m2} \approx \delta t(\mu_d - \mu_h) \ + \ s_{m1}$$

The effective selection coefficient in this case has two terms. The first term gives the strength of selection on the new allele due to differences in the baseline mortality rates between the phases. The second term is the component of selection due to selection against the mutation. and is identical to equation 5, the effective selection coefficient on new life cycle alleles from the mutation only model.

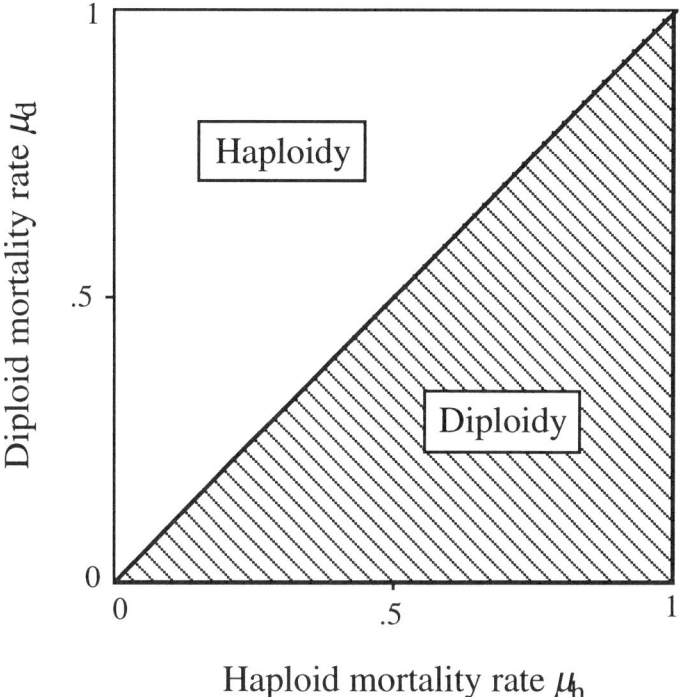

FIGURE 4. Stable life cycles under mutation and ecological selection. The striped area indicates parameters for which alleles that increase diploidy are favored and the white area indicates that increased haploidy is favored. For points along the diagonal line, the results from Model 1 will hold (Fig. 2).

As indicated in the results from Model 1, the magnitude of the selection coefficient on the new life cycle allele due only to selection against the mutation (s_{ml}) is on the order of the per-locus mutation rate, u. We can therefore see from equation 8 that if $\| \mu_d - \mu_h \| \gg u$, the ecological component of selection will be a stronger evolutionary force than the selection against the mutation. Life cycles that result are shown in Figure 4. Regardless of the amount of recombination or the values of the mutant allele parameters, diploidy is favored if the diploid phase has a lower baseline mortality rate than haploid phase. Conversely, when the haploid mortality rate is lower than the diploid, alleles that increase the haploid phase are favored. In each of these cases, the life cycle that maximizes mean fitness is stable. The strength of selection on the life cycle alleles is approximately equal to $-\delta t\,(\mu_d - \mu_h)$, the difference between the mortality rates times the size of the life cycle change from the new allele.

If, on the other hand, $\| \mu_d - \mu_h \| \ll u,$ the second term of equation 8 will control the behavior of the system and life cycles will evolve as described in Model 1. The range of parameter space for which this will be true is approximately represented by the thickness of the diagonal line shown in Figure 4.

Under the model of ecological selection used here, where mortality rates per unit time are constant, either haploidy or diploidy is the only stable life cycle. Once again, life cycles that include both phases ($0 < t < 1$) are not stable.

Discussion

The effect of selection against deleterious mutations on the evolution of life cycles. The central conclusion of the analysis of Model 1 is that it is easy to have stable diploid life cycles when selection is weak, mutations are partially recessive and there is some level of recombination in the genome. There is strong empirical evidence that most deleterious mutations are partially rather than fully recessive (Simmons & Crow, 1977), even when they appear in organisms where most of the selection takes place in the haploid phase (Orr, 1991). There is also evidence that the dominance of new mutations is weaker the stronger the selection against them (Charlesworth, 1979). Given that recombination at some positive level occurs in most organisms, we should expect to see evolution toward life cycles with longer diploid phases under these biologically reasonable conditions.

This conclusion is supported by the single-locus analysis in Model 1 above and qualitatively consistent with the results from Perrot, Richerd & Valero (1991), Bengtsson (1992), Otto & Goldstein (1992) and Otto (this volume). It is also supported by a study undertaken by the authors of mutations at a large number of loci throughout the genome (Jenkins & Kirkpatrick, unpublished manuscript). If selection and mutation rates are identical at each locus, and the loci act independently (fitnesses are multiplicative), life cycles evolve exactly as in Model 1, with results as shown in Figure 2. Epistasis is not necessary in order for diploid life cycles to evolve. Therefore, it appears that the conflict in results reached by previous theoretical studies is not due to differences between single- and multi-locus approaches but to the different methods of analysis used.

It is clear from our results that genes that control life cycles do not necessarily evolve to maximize population mean fitness. In any situation where group selection operates, however, the results from Kondrashov & Crow (1991) will be valid; only haploidy will be stable unless there are strong interactions between loci across the genome. Group selection arguments are obviously appropriate for non-interbreeding populations with different life cycles. They are also appropriate, though not so obviously, in two additional cases. In a life cycle that has synchronous meiosis rather than synchronous syngamy, the haploid genotype at the life cycle locus controls the life cycle and assortative mating of life cycle types would occur. If assortment is complete, this effectively results in reproductively isolated groups and the Kondrashov & Crow (1991) results hold. In addition, as indicated in the analysis of Model 1, the life cycle that maximizes mean fitness does evolve when there is complete linkage between the life cycle and viability locus. Such a "congealed" genome may have existed early in the evolution of eukaryotic organisms if meiotic reproduction preceded recombination. Therefore, mean fitness arguments may be important for understanding the origin of eukaryotic life cycles and the origin of sexual reproduction itself.

The conclusions drawn from Model 1 above and from other single-locus models like it (Perrot, Richerd & Valero, 1991; Bengtsson, 1992; Otto & Goldstein, 1992; Otto, this volume) must be tempered by a strong caveat. The evolutionary force on the modifier of the life cycle is very weak; the strength of selection on a life cycle allele is on the order of the per locus mutation rate, likely to be somewhere from 10^{-4} to 10^{-6} per locus per generation (Kondrashov, 1988; Drake, 1991). This level of selection could easily be overwhelmed by other evolutionary forces, such as pleiotropic selection on the life cycle modifier or differing ecological selection on the life cycle phases (as in Model 2). Given, however, that deleterious mutations occur at loci across the genome, the effects of this type of evolutionary force might be substantial if cumulative effects of mutations throughout the genome become appreciable. Results from our study of mutations at a large number of loci indicate that the strength of selection on modifiers of the life cycle then becomes of the order of the genome-wide mutation rate (Jenkins & Kirkpatrick, unpublished manuscript). Empirical evidence suggests that this rate may be greater than 0.1 new mutations per genome per generation (Simmons & Crow, 1977; Kondrashov, 1988; Houle *et.al.*, 1992). Mutation rates of this size could be of evolutionary importance.

The effect of ecological selection and comparison of effects. The main conclusion from our analysis of Model 2 is that, if both selection against mutations and ecological selection are operating, the stronger of the two forces determines the evolutionary outcome. If the difference between the baseline mortality rates in the two phases is much larger than the per-locus mutation rate, the ecological component of selection will be a stronger evolutionary force. In this case, the phase with the lower mortality rate is favored. If, on the other hand, the difference between the haploid and diploid baseline mortality rates is less than the per-locus mutation rate, selection against the mutation is the stronger evolutionary force and life cycles will evolve as described in Model 1. In this case, diploidy may often be favored under biologically reasonable conditions.

Though this result is not surprising, it raises a question as to the power of selection against deleterious mutations to explain the evolution of life cycles if other types of selection are also operating. One consideration is the possible effect of all forces across the entire genome. Deleterious mutations certainly occur at all loci in both phases, and as discussed above, the cumulative effect of selection against them may lead to a substantial evolutionary force on life cycles (Jenkins & Kirkpatrick, unpublished manuscript). We have little empirical information as to the effect of other types of selection, however, so it is difficult to determine their importance. For example, ecological selection may favor the haploid phase for some traits and the diploid phase for others, leading to small net selection on the life cycle. Whenever this is the case, the force of selection against deleterious mutations may be evolutionarily important.

It is clear from the analysis of Model 1 that selection against deleterious mutations is not sufficient to explain the evolution of all life cycles seen in nature, for stable life cycles that retain both phases are not supported by this model. As shown in Model 2 and previous work (Jenkins, 1993), constant mortality rates that differ between the haploid and diploid phase also will not result in mixed life cycles. This suggests that other factors, such as the different forms of ecological selection discussed by Jenkins (1993), may be important in the evolution of such life cycles. Further theoretical treatment of the evolutionary effect of other types of ecological selection is needed.

There is also a need for a better understanding of the types and strengths of selection acting on haploid and diploid phases. Many suggestions have been made as to possible ecological and physiological differences between the phases (Cavalier-Smith, 1978; Mulcahy, 1979; Willson, 1981; Lewis, 1985; Buss, 1987; Cousens 1988; De Wreede & Klinger, 1988; McLachlan, 1991) but little empirical information is available (see Perrot, this volume). Recent field studies (Destombe *et al.*, 1989) and experimental phase-specific selection studies (Destombe *et al.*, 1993) begin to address these needs, but far more information concerning how selection operates on haploid and diploid phases is required.

REFERENCES

1. BELL, G., The comparative biology of the alternation of generations, this volume.

2. BENGTSSON, B. O., Deleterious mutations and the origin of the meiotic ploidy cycle, Genetics **131** (1992), 741-744.

3. BERNSTEIN, H., G. S. BYERS and R. E. MICHOD, Evolution of sexual reproduction: importance of DNA repair, complementation, and variation, Am. Nat. **117** (1981), 537-549.

4. BUSS, L. W., *The Evolution of Individuality*, Princeton University Press, Princeton, (1987).

5. CAVALIER-SMITH, T., Nuclear volume control by nucleoskeletal DNA, selection for cell volume and cell growth rate, and the selection of the DNA C-value paradox, J. Cell Sci. **34** (1978), 247-278.

6. CHARLESWORTH, B., Evidence against Fisher's theory of dominance, Nature **278** (1979), 848-849.

7. COUSENS, M. I., Reproductive strategies of pteridophytes, in *Plant Reproductive Ecology: Patterns and strategies,* edited by J. Lovett Doust and L. Lovett Doust. Oxford University Press, New York, (1988), 307-328.

8. CROW, J. F. and M. KIMURA, Evolution in sexual and asexual populations, Am. Nat. **99** (1965), 439-450.

9. CROW, J. F. and M. KIMURA, *An Introduction to Population Genetics Theory,* Harper & Row, New York, (1970).

10. DESTOMBE, C., M. VALERO, P. VERNET and D. COUVEt, What controls haploid-diploid ratio in the red alga, *Gracilaria verrucosa*? J. Evol. Biol. **2** (1989), 317-338.

11. DESTOMBE, C., J. GODIN, M. NOCHER, S. RICHERD and M. VALERO, Differences in responses between haploid and diploid isomorphic phases of *Gracilaria verrucosa* (Rhodophyta: Gigartinales) exposed to artificial environmental conditions, Hydrobiologia (in press, 1993).

12. DE WREEDE, R. E. and T. KLINGEr, Reproductive strategies in alga in *Plant Reproductive Ecology: Patterns and strategies,* edited by J. Lovett Doust and L. Lovett Doust. Oxford University Press, New York, (1988), 267-284.

13. DRAKE, J. W., Spontaneous mutation, Ann. Rev. Genet. **25** (1991), 125-146.

14. HOULE, D., D. K. HOFFMASTER, S. ASSIMACOPOULOS and B. CHARLESWORTH, The genomic mutation rate for fitness in *Drosophilia*, Nature **359** (1992), 58-60.

15. JENKINS, C. D., Selection and the evolution of genetic life cycles, Genetics **133** (1993), 401-410.

16. JENKINS, C. D. and M. KIRKPATRICK, Deleterious mutations and the evolution of genetic life cycles (unpublished manuscript).

17. KONDRASHOV, A. S., Deleterious mutations and the evolution of sexual reproduction, Nature **336** (1988), 435-440.

18. KONDRASHOV, A. S. and J. F. CROW, Haploidy or diploidy: which is better? Nature **351** (1991), 314-315.

19. LEWIS, W. M., Nutrient scarcity as an evolutionary cause of haploidy, Am. Nat. **125** (1985), 692-701.

20. MCLACHLAN, J. L., *Chondrus crispus* (Irish moss), an ecologically important and commercially valuable species of red seaweed of the North Atlantic Ocean, in *Marine Biology: Its Accomplishments and Future Prospects,* edited by J. Mauchline and T. Nemoto. Hokusen-sha Publishing Co., Tokyo (1991), 217-233.

21. MULCAHY, D. L., The rise of the angiosperms: a genecological factor, Science **206** (1979), 20-23.

22. MULLER, H. J., Some genetic aspects of sex, Am. Nat. **66** (1932), 118-138.

23. ORR, H. A., A test of Fisher's theory of dominance, Proc. Natl. Acad. Sci. USA, **88** (1991), 11413-11415.

24. OTTO, S., The role of deleterious and beneficial mutations in the evolution of ploidy levels, this volume.

25. OTTO, S. and D. GOLDSTEIN, Recombination and the evolution of diploidy, Genetics **131** (1992), 745-751.

26. PERROT, V., Empirical approaches to the evolution of life cycles, this volume.

27. PERROT, V., S. RICHERD and M. VALERO, Transition from haploidy to diploidy, Nature **351** (1991), 315-317.

28. SIMMONS, M. J. and J. F. CROW, Mutations affecting fitness in Drosophila populations, Ann. Rev. Genet. **11** (1977), 49-78.

29. VALERO, M., S. RICHERD, V. PERROT and C. DESTOMBE, Evolution of alternation of haploid and diploid phases in life cycles, Trends. Ecol. Evol. **7** (1992), 25-29.

30. WILLSON, M. F., On the evolution of complex life cycles in plants: a review and an ecological perspective, Ann. Missouri Bot. Gard. **68** (1981), 275-300.

DEPARTMENT OF ZOOLOGY
UNIVERSITY OF TEXAS AT AUSTIN
AUSTIN, TEXAS 78712

jenkins@biff.zo.utexas.edu
kirkp@biff.zo.utexas.edu

Lectures on Mathematics in the Life Sciences
Volume **25**, 1994

The Role of Deleterious and Beneficial
Mutations in the Evolution of Ploidy Levels

SARAH P. OTTO

October 8, 1993

ABSTRACT. All sexual organisms experience a haploid and a diploid phase during their life cycle, yet the extent of each phase is remarkably variable. We explore the evolution of a locus that alters the timing of meiosis, hence altering the proportion of the life cycle spent in either the haploid or diploid phase. Evolution of the life cycle will occur in response to viability selection acting at fitness loci even when selection does not directly act on the locus modifying the life cycle. Both selection against deleterious mutations and for beneficial mutations are considered. It is found that even when diploids mask deleterious alleles and reveal beneficial ones, the diploid phase of the life cycle will not necessarily increase. Conditions under which each phase is expected to increase are developed. These theoretical results may, in part, explain observed life cycle variation.

1. Introduction

Nearly a century has passed since the recognition of chromatin as the hereditary material underlying evolution (elucidated in parts by Weismann, Hertwig, Kölloker, Strasburger, Sutton, Boveri, and others [**29**]). In 1883, van Beneden found that the number of chromosomes halved during the production of gametes only to double again during fertilization in *Ascaris bivalens*, a threadworm [**29**], thus showing that the number of chromosomes does not remain constant throughout a life cycle. Cytological studies in the 1890's by Strasburger, Guignard, Overton, Farmer and others [**40**] extended this work to various species of plants. These studies revealed the relationship between chromosome number and the alternation of generations, with a reduction division (meiosis) generally

1991 *Mathematics Subject Classification*. Primary 92D15; Secondary 65C20, 15A42.

The author was supported by a Miller Post-Doctoral Fellowship and by NIH grant number GM40282 to Montgomery Slatkin.

The final version of this paper will be submitted for publication elsewhere.

leading to the gametophytic (haploid or x) stage and syngamy leading to the sporophytic (diploid or $2x$) stage. By the early part of the twentieth century, it was understood that some organisms are dominated by the x stage (examples among the green algae), some are dominated by the $2x$ stage (gymnosperms and angiosperms) and yet others are characterized by the extensive development of both $2x$ and x tissue (bryophytes, ferns) [28, 40]. This information was immediately placed in an evolutionary context, as exemplified in the following quote:

> In the vegetable kingdom evolution seems to have been accompanied by a gradual increase of the $2x$-generation, and a corresponding reduction of the x-generation in point of importance.
> – R. H. Lock, 1906, p. 271.

Perhaps typically, Lock saw the increase in diploidy as he surveyed from the "lowly marine organisms and passing upwards... to the flowering plants," as evidence for an advantage to diploidy. He identified a possible advantage emanating from the genotypic variability produced by the union of two distinct genomes within an individual:

> ...it is only in [$2x$ organisms] that the operation of Mendel's law can lead to the production of new combinations of parental characters in the body which represents the main stage of the life history; and that this circumstance may possibly lead to a greater power of adaptability to external circumstances.
> – R. H. Lock, 1906, p. 275.

During this century, many have followed Lock in a belief that diploids tend to be more genetically variable and are consequently favored by evolution [3, 7, 8, 41]. One form of this argument is that diploids are able to combine alleles into the heterozygous condition and thus have a broader range of genotypes to explore. Another argument is that diplonts (organisms with syngamy immediately after meiosis) produce genetically diverse gametes whereas haplonts (organisms with meiosis immediately after syngamy) produce genetically identical gametes and this might enhance the competitive advantage of diploid life cycles when variability is favorable [3]. Lastly, it has been thought that deleterious recessive mutations, maintained at a higher frequency in diploids, can be an important source of genetic variation in the face of new environmental conditions [22, 37, 38].

Others have focused directly on the selective consequences to an individual of its ploidy level, rather than focusing on the extent of variation found in haploids or diploids. For example, selection occasionally favors heterozygotes and this type of selection will, to the extent of its prevalence, favor diploidy [4, 6, 16]. Another aspect of diploidy that is thought to be to its favor is that rare deleterious mutations will almost always be accompanied by normal alleles in a diploid individual [38]. These normal alleles frequently compensate for a good portion of the deleterious effects of a mutation [39], which are then said to be masked. This masking of mutations has generally been thought to favor diploidy [20, 36].

Complementing these theories that overdominant selection and selection against deleterious mutations favor diploidy, Paquin and Adams argued that selection for beneficial mutations would favor diploids, since diploids have twice as many genes which can mutate to new favorable alleles [34].

There is a tendency among evolutionary biologists to use complexity as an appraisal of the evolutionary status of an organism [37]. Thus, much attention has been focused on finding the evolutionary advantage possessed by diploids over haploids. The persistence and ubiquity of haploid forms, which overwhelm diploids in sheer number, belies this focus on the evolution of diploidy. The real task ahead is to identify the factors that favor changes in the ploidy level of an organism (whether by extending the haploid or diploid phase) and those that favor the maintenance of a given ploidy level. Why is it that we see closely related species of algae [3] and yeast [14] with dramatically different life cycles? Why are there organisms that maintain both a haploid and a diploid phase to their life cycles in the face of variants that would allow dominance of one or the other phase [9, 10, 21]? While more experimental work is keenly needed, we are also in need of theoretical guidance. Only recently have models addressed the masking hypothesis [27, 32, 33, 36], the overdominance hypothesis [16], and hypotheses related to the maintenance of both haploidy and diploidy [23]. The results have often been surprising, with common wisdom failing under scrutiny. Masking has been found to favor *haploidy* under conditions of low genetic mixing within a population [32, 33]. Selection with one allele favored in the haploid phase and a different allele favored in the diploid phase does not support the maintenance of both phases within a life cycle unless special conditions hold[23].

In this work, it will be shown that deleterious and beneficial mutations can impart an advantage to diploidy *or to haploidy*. Specific conditions are outlined under which we expect an increase in one or the other phase of a life cycle. To obtain these conditions, we use a model that closely mimics known life cycles, with individuals passing through and experiencing selection both as haploids and as diploids. The results for deleterious mutations are consistent with those obtained by Otto and Goldstein [32] for a different life cycle. We then present the first theoretical analysis of the effects of beneficial mutations on ploidy levels. With beneficial mutations, diploids are more likely to bear new advantageous alleles. Selection is more effective, however, in haploids, so that beneficial mutations tend to sweep faster through the more haploid members of a population, imparting an advantage upon haploidy. Which effect is more important depends on the specific selective and genetic conditions under consideration.

2. Methods

We wish to understand the evolution of ploidy levels. In wholly or partially asexual populations, the evolution of sex will necessarily entail changes in ploidy levels of the population at certain stages (see articles by Michod and Gayley and

by Kondrashov in this collection). In this paper, however, we wish to isolate selection for haploidy or diploidy from selection for sexual or asexual reproduction.
We thus assume a sexual life cycle or at least a life cycle with a constant proportion of sexuality to asexuality. We then postulate the existence of a gene (the
ploidy locus, C), the effect of which is to hasten or delay the timing of meiosis
relative to that of syngamy and zygote formation (Figure 1, [**23**]). The proportion of time spent in each phase of the life cycle depends on the genotype (C_iC_j)
of the diploid zygote at this locus, according to Table 1. Genetic changes at this
ploidy locus can modify the life cycle by either a small amount (micromutations)
or dramatically; for instance, it is possible that a single mutation changes an organism from a haplont to a diplont. For mathematical convenience, we assume
non-overlapping generations with all individuals in the population producing gametes at the same time regardless of their genotype. Changes in the frequency
of an allele at the ploidy locus are assumed to occur indirectly, in response to the
selective regime acting at the second locus (the viability locus, A). The survival
rate of an individual per generation (or the fertility rate, assuming individual-
and not couple-dependent fertility selection) depends, by assumption, only on
its genotype at the viability locus according to Table 2. For any particular
chromosome, the survival rate per generation must be adjusted according to the
proportion of a generation spent in the diploid phase (d_{ij}) and in the haploid
phase ($1 - d_{ij}$). For instance, consider an aC_1 chromosome within an AaC_1C_1
diploid: the probability that it will survive through the diploid phase of selection is $(1 - hs)^{d_{11}}$; the haploid aC_1 offspring then produced will survive with
probability $(1 - s)^{1 - d_{11}}$. Notice that, if d_{ij} equals one, nascent C_iC_j individuals
experience selection only as diploids (or as haploids if $d_{ij} = 0$). The ploidy and
viability loci recombine at meiosis with a probability of r ($0 \leq r \leq 1/2$). Mutations occur at a rate, $\mu_1/2$, from the wildtype (A) allele to the mutant allele
(a) at the viability locus during meiosis and at a rate of $\mu_2/2$ during gametogenesis (see Figure 1; $\mu_1 = \mu_2 \equiv \mu$ will generally be assumed where the total
mutation rate per cycle is μ). The results do depend quantitatively on the time
during the life cycle when mutations occur[1] (especially when s is large); the
qualitative results that will be discussed, however, are independent of the place-
ment of mutation. The mutation process described allows mutations to occur
during the reproduction of both haploids and diploids; this not only appears to
be more realistic but also provides results that are intermediate between those
when mutation is placed at meiosis or at syngamy.

We first examine the effects of deleterious mutations (mutant fitnesses less
than one) on evolution at the ploidy locus and then the effects of beneficial
mutations (mutant fitnesses greater than one).

---

[1]For example, Jenkins and Kirkpatrick, this volume, place mutation at meiosis only and
obtain slightly different results.

Genotype	Proportion of time spent as a	
	Diploid	Haploid
$C_1 C_1$	d_{11}	$1 - d_{11}$
$C_1 C_2$	d_{12}	$1 - d_{12}$
$C_2 C_2$	d_{22}	$1 - d_{22}$

TABLE 1. The determination of ploidy level.

Genotype	Viability
AA	1
Aa	$(1 \pm hs)^{d_{ij}}$
aa	$(1 \pm s)^{d_{ij}}$
A	1
a	$(1 \pm s)^{d_{ij}}$

TABLE 2. Viability selection at the A/a locus. Selection is positive (+) for beneficial mutations and negative (−) for deleterious ones.

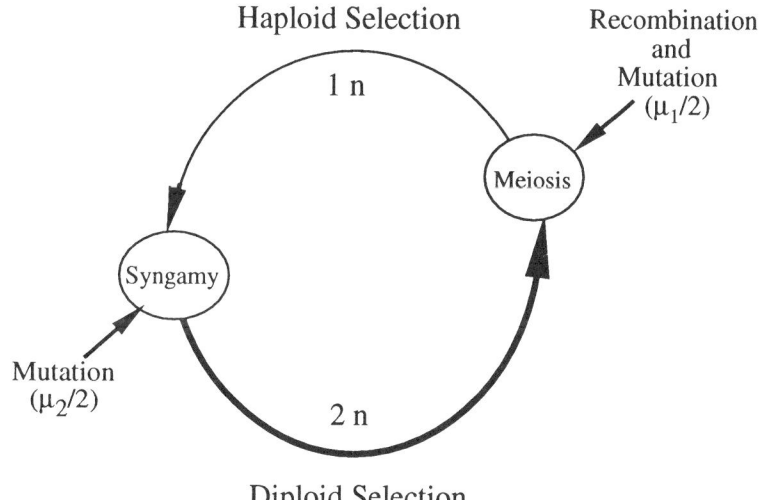

FIGURE 1. Life cycle with alternation of generations. Selection occurs in both the haploid and the diploid phase. The extent of selection in each phase depends on the genotype at a locus (the ploidy locus) which controls the timing of meiosis. Meiosis may occur immediately following syngamy, immediately before it, or any time in between.

3. Results

Following the chromosomal frequencies from one generation to the next, the recursions given in the Appendix A may be derived. These recursions form the basis for the following results.

3.1. Deleterious Mutations. Examining models of individual selection, Perrot, Richerd and Valero [36] observed that the masking of deleterious mutations would favor the evolution of diploidy, assuming random mating and free recombination between the loci. Otto and Goldstein [32] pointed out that masking has the second effect of allowing mutations to persist and reach higher frequencies in diploids (see also [15] and [33]). Under conditions of low genetic mixing within a population (low recombination, frequent assortative mating, selfing or asexual reproduction), the chance that a diploid will carry a mutation is so much higher than the probability that a haploid will carry one that individual selection favors the evolution of haploidy, even though diploids mask the mutations that they do carry.

With selection against the a allele ($s > 0$) and with the C_1 allele fixed at the ploidy locus, the recursions in Appendix A approach a polymorphic equilibrium where mutation and selection are balanced:

$$(3.1) \qquad \frac{\mu[1 + (1-s)^{(1-d_{11})}]}{2[1 - (1-s)^{(1-d_{11})}(1-hs)^{(1-d_{11})}]}$$

A new ploidy allele increases in frequency from this equilibrium if the leading eigenvalue determined from the recursions is greater than one (see Appendix B). The leading eigenvalue, equation (B.2), depends on all of the variables of the model in a complicated manner. A few observations may be drawn from a close examination of the leading eigenvalue:

- Whether the leading eigenvalue is greater than or less than unity does not depend on the mutation rate, but the extent of the departure from one is proportional to μ.
- The value of the eigenvalue increases monotonically with r when the new ploidy allele increases the extent of diploid selection.
- The value of the eigenvalue decreases monotonically with r when the new ploidy allele increases the extent of haploid selection.
- When $r = 0$, haploidy is always favored.
- As r is increased from zero, the parameter space in which diploidy is favored increases, with diploidy being favored most often when mutations are highly recessive (h near zero), as illustrated in Figure 2.

Each of these results is completely analogous to results derived in [32]. Thus we see that linkage favors the evolution of haploidy while extensive recombination favors the evolution of diploidy. The major departure from previous results is that the invasion criterion is now sensitive to the extent of diploidy among the resident population (d_{11}), as illustrated in Figures 3-5. Above each curve are

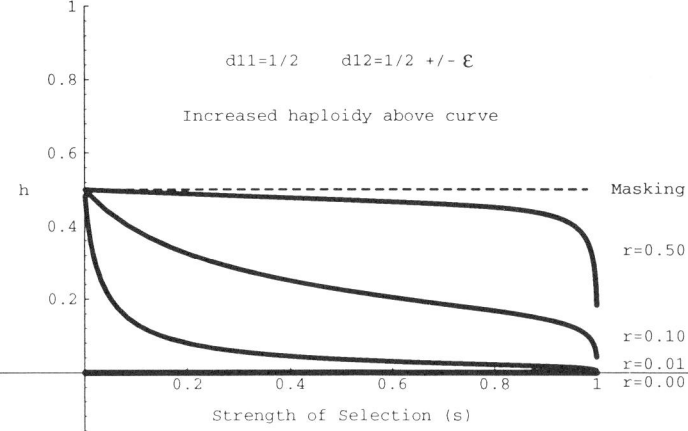

d11=1/2   d12=1/2 +/- ϵ

Increased haploidy above curve

Masking

r=0.50

r=0.10
r=0.01
r=0.00

Strength of Selection (s)

FIGURE 2. Parameters favoring the evolution of increased haploidy (above curve) or increased diploidy (below curve) for various recombination rates. Note that masking occurs in the entire region below the dashed line ($h < 1/2$). Here and elsewhere, ϵ refers to a small quantity.

the values of h and s for which haploidy is favored; below each curve diploidy is favored. The results are clearly sensitive to the exact ploidy alleles under consideration. Consider the point marked by a $*$ ($h = 0.4, s = 0.9$) with free recombination. An initially haploid population ($d_{11} = 0$, Figure 3) cannot be invaded by a ploidy allele that increases the diploid phase by a small amount ($d_{12} = \epsilon$), or by an intermediate amount ($d_{12} = 1/2$), but can be invaded by an allele that leads to a large diploid phase ($d_{12} = 1$). An initial population that experiences both phases equally ($d_{11} = 1/2$, Figure 4) can be invaded by alleles that cause either complete diploidy ($d_{12} = 1$) or complete haploid ($d_{12} = 0$), but can only be invaded by micromutations that increase the extent of diploidy ($d_{12} = 1/2 + \epsilon$). An initially diploid population ($d_{11} = 1$, Figure 5) cannot be invaded by any alleles. Graphical analyses indicate that populations with intermediate values of d_{ij} are more susceptible to invasion and that the initial increase of extreme values of d_{ij} (near zero or one) is supported over a larger parameter range than the invasion of intermediate values. In short, alleles that maintain both phases of the life cycle in intermediate amounts are evolutionarily unstable in this model (in the EGS sense [11]); the dominance of either the haploid phase *or* the diploid phase is expected given the appearance of sufficiently diverse modifier alleles. These results thus shed no light on the maintenance of alternating generations, but rather make its occurrence more puzzling.

Despite the quantitative differences between the results obtained using the life cycle illustrated in Figure 1 and that used by Otto and Goldstein [32], the qualitative result that genetic associations develop which favor haploidy when

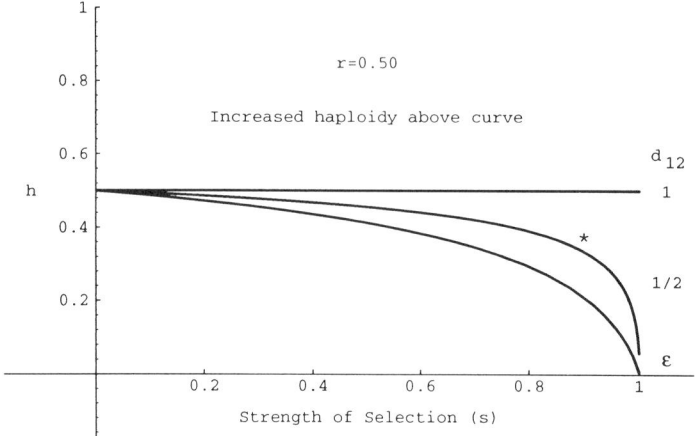

FIGURE 3. Sensitivity to initial ploidy level: starting population is haploid, $d_{11}=0$. For different values of d_{12}, the curve is given below which the rare allele with the new ploidy level can invade, thus increasing the extent of the diploid phase. Above the curves, the resident ploidy allele is stable to invasion. An $*$ is placed where $s = 0.9$ and $h = 0.375$ for discussion within the text.

recombination rates are low is common to both. These associations develop because individuals bearing mutations are more likely to die if they are haploid since the mutation is then unmasked. Those haploid individuals that do not die but survive selection are less likely to carry mutations and are less likely to bear offspring with mutations. From the opposite perspective, masking permits diploids carrying mutations to survive selection but it thereby allows those mutations to persist among the offspring of these diploids. Consequently, the frequency of deleterious mutations becomes higher among individuals with longer diploid phases than among individuals with longer haploid phases, an association that favors haploidy.

3.1.1. *Comments about non-random mating.* Otto and Marks [33] found that selfing, assortative mating, parthenogenesis and other forms of asexual reproduction favor the evolution of haploidy by limiting the genetic mixing that separate haploid (diploid) ploidy alleles from the viability alleles that have recently experienced haploid (diploid) selection. I investigated these forms of non-random mating for the life cycle currently under analysis, finding once again that their effect was to strengthen the extent of genetic associations thereby increasing the parameter range in which haploidy was favored. For any particular organism, mating patterns must thus be understood before we can predict whether haploidy or diploidy would be favored by selection against deleterious mutations.

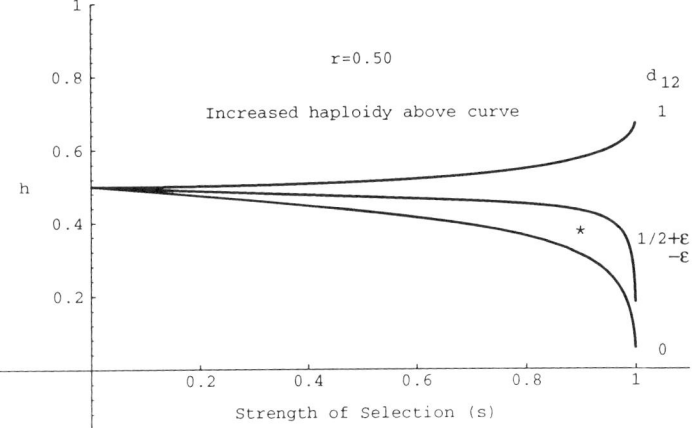

FIGURE 4. Sensitivity to initial ploidy level: starting population is haplont-diplont, $d_{11}=1/2$. For different values of d_{12}, the curve is given below which the rare ploidy allele can invade if $d_{12} > d_{11}$ but not if $d_{12} < d_{11}$. Above the curves, the new ploidy allele can invade if $d_{12} < d_{11}$ but not if $d_{12} > d_{11}$. An $*$ is placed where $s = 0.9$ and $h = 0.375$ for discussion within the text.

3.2. Beneficial Mutations. Paquin and Adams [**34**] argued that the number of beneficial mutations experienced by a diploid population would be larger than that of a haploid population because twice as many genes would be able to mutate. Their arguments were based on several assumptions: that the populations of haploids and diploids do not interbreed, that the number of haploids would be less than twice the number of diploids, and that there would be no significant differences in the probability that a beneficial mutation would become fixed within a haploid versus a diploid population. We evaluate the influence of sweeps of beneficial mutations on the evolution of ploidy levels in the context described by Paquin and Adams, with non-interbreeding asexual populations, and then in the context of an interbreeding population of sexual haploids and diploids as in Figure 1.

3.2.1. *Asexual haploid and diploid populations.* We wish to know when the rate of adaptation, defined as the rate of accumulation of beneficial mutations, will be faster in a diploid asexual population than in a separate haploid one. While diploids have twice as many mutations within their doubled genome, these new mutations arise in the heterozygous state and are generally less advantageous than if they arose unmasked in a haploid. In comparing the rate of adaptation of separate haploid and diploid populations, then, we must consider both the increased number of mutations in diploids and the decreased advantage of these mutations.

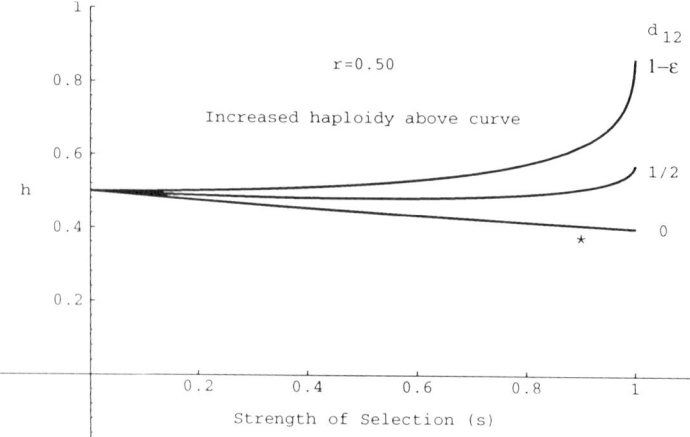

FIGURE 5. Sensitivity to initial ploidy level: starting population is diploid, $d_{11}=1$. For different values of d_{12}, the curve is given above which the rare allele with the new ploidy level can invade, thus increasing the extent of the haploid phase. Below the curves, the resident ploidy allele is stable to invasion. An $*$ is placed where $s = 0.9$ and $h = 0.375$ for discussion within the text.

In asexual populations, beneficial alleles can only accumulate within a genome if successive mutations occur within a single lineage; mutations occurring in different individuals cannot, in the absence of sex, be recombined into one individual [30]. If mutations are infrequent such that a mutation arises and fixes before the next beneficial mutation occurs, then these new mutations will always occur in the lineage with all previous beneficial mutations (technically, all those that survived loss when rare). Here, the nesting of beneficial mutations is ensured. On the other hand, if mutations are frequent, new mutations can occur in different lineages that are present concurrently, but only one of these lineages will ultimately survive. All new mutations that do not occur within the "lucky" lineage will ultimately be lost, even though they impart a selective advantage to their carriers. As a concrete example, consider a new mutation that occurs within a population with a previous beneficial mutation segregating at a frequency of 5%. Most likely, this new mutation will not occur within an individual bearing the previous mutation; all individuals will thus carry no mutations or a single mutation. The single mutant with the highest fitness advantage will eventually fix and the other will be lost from the population. If mutations are very frequent, then, many mutations will simply be lost by competitive exclusion. In this case, the rate of adaptation is generally limited not by the rate of mutation, but by the rate of spread of beneficial alleles (that is, by their selective advantage). The faster good alleles spread, the sooner new mutations can be nested within the same lineage. In short, the rate of fixation of beneficial mutations in asexual populations will tend to be limited by one of two factors: the mutation rate to new advantageous alleles (when mutations are rare) or the time required for beneficial mutants to reach appreciable frequencies (when mutations are frequent).

In this section, we will determine quantitatively how the rate of adaptation depends on mutation rates and selection coefficients. These results will be discussed in light of expected differences between haploid and diploid asexual populations. We will then reexamine the data of Paquin and Adams with *Saccharomyces cerevisiae*, from which they infer more sweeps of beneficial mutations in diploid populations.

The following analysis is similar in structure to that presented by Kirkpatrick and Jenkins [26] and by Weiner et al. [43]. Consider an asexual population of size N with beneficial mutations occurring at a genome-wide rate equal to U. *If* no other mutations are segregating within the population, a new beneficial mutation that arises will have a probability of eventually becoming fixed that is approximately equal to 2σ, where $1 + \sigma$ is the relative fitness of an individual carrying this new mutation and is assumed to be near one [19]. Thus the rate that new beneficial mutations (destined to fix) will arise in an asexual population is:

$$(3.2) \qquad \rho_m = N \, U \, 2\sigma.$$

We may use equation (3.2) to compare the rate of adaptation of separate haploid and diploid populations. Let N_H and N_D be the haploid and diploid population sizes, respectively. Assume that beneficial mutations occur at a rate μ per gene per generation, and that there are L genes that may experience beneficial mutations in haploids but twice as many, $2L$, in diploids ($U = \mu L$ for haploids, but $2\mu L$ for diploids). Further, let the relative fitness of an individual with a new beneficial mutation be $1 + s$ in haploids and $1 + hs$ in diploids. Realistically, s and h are random variables following some unknown distribution; for simplicity, we will assume that s and h are constant. We will also assume that h tends to be less than one, presuming that an advantageous allele that occurs within an individual carrying another allele at the same locus (i. e. a diploid) will be less effective. Equation (3.2) thus becomes:

$$(3.3) \qquad \rho_m^{hap} = N_H \; \mu L \; 2s \qquad \text{for the haploid population and}$$
$$(3.4) \qquad \rho_m^{dip} = N_D \; 2\mu L \; 2hs \qquad \text{for the diploid population,}$$

which assume that each beneficial mutation sweeps through the population before the next mutation appears. For haploid and diploid populations of equivalent size, the rate of adaptation in a diploid population will be $2h$ times that of a haploid population. The diploid population will fix dominant mutations ($h > 1/2$) at a faster rate, but will accumulate beneficial recessive changes ($h < 1/2$) at a slower rate.

When beneficial mutations arise frequently, as they would in very large populations, the probability that any new mutation will survive to fix in the population is reduced from the value of 2σ, since these mutations may arise in individuals that do not have the highest fitness within the population. The faster beneficial alleles sweep through the population, however, the more likely it is that the fittest genotype will be at a high frequency, and the more likely new mutations will be nested within this lineage. To tell whether beneficial mutations tend to occur while previous mutations are still segregating, we now give the time required for a beneficial mutation to fix. In an asexual population of size N, a beneficial mutation that arises with frequency p_1 and selective advantage σ will reach fixation in $\bar{t}_f(p_1)$ generations, on average [25], where

$$(3.5) \quad \bar{t}_f(p_1) \approx \frac{1}{\sigma(1 - e^{-2N\sigma})} \int_{p_1}^1 \frac{(e^{2N\sigma x} - 1)(e^{-2N\sigma x} - e^{-2N\sigma})}{x(1 - x)} dx$$

if p_1 is very small (to translate the model used by Kimura and Ohta into this asexual model, their selection coefficient was doubled and their population size halved). For N sufficiently large, $\bar{t}_f(p_1)$ may be approximated by the deterministic analog calculated as in section 2.1 of [12],

$$
\begin{aligned}
(3.6) \quad \bar{t}_f(p_1, p_2) &\approx \int_{p_1}^{p_2} \frac{1 + \sigma x}{\sigma x(1 - x)} dx \\
&= \frac{1}{\sigma}(ln[p_2(p_2 - 1)^{-(1+\sigma)}] - ln[p_1(p_1 - 1)^{-(1+\sigma)}]), \\
&= \frac{1}{\sigma}(ln[\frac{p_2}{p_1}] - (1 + \sigma)ln[\frac{1 - p_2}{1 - p_1}]),
\end{aligned}
$$

where p_2 is the gene frequency assigned as the fixation point (recall that an infinite population would take an infinite amount of time to fix completely). If we measure the time that an allele currently in one individual would be in every individual of the population except one ($p_1 = 1/N$ and $p_2 = 1 - 1/N$), equation (3.6) becomes:

$$(3.7) \qquad \bar{t}_f(1/N, 1 - 1/N) \approx \frac{(2 + \sigma)}{\sigma} ln[N - 1].$$

For large N and small σ, we see from equation (3.7) that the time to fixation is roughly proportional to $1/\sigma$ (with the rate of fixation, $\rho_f \equiv 1/\bar{t}_f$, being proportional to σ). In short, if the rate of adaptation is limited not by the appearance of mutations but rather by their spread, we might expect that the rate of accumulation of beneficial mutations would be proportional to the selective advantage of the alleles.

Applying these results to a comparison of haploid and diploid populations, let σ equal s for haploid mutants and hs for diploid mutants, as before. Using equation (3.7), we see that the rates at which beneficial alleles fix in haploid and diploid populations, respectively, are:

$$(3.8) \qquad \rho_f^{hap} \equiv \frac{1}{\bar{t}_f^{\,hap}(1/N_H, 1-1/N_H)} \quad \approx \quad \frac{s}{(2+s)ln[N_H-1]},$$

$$(3.9) \qquad \rho_f^{dip} \equiv \frac{1}{\bar{t}_f^{\,dip}(1/N_D, 1-1/N_D)} \quad \approx \quad \frac{hs}{(2+hs)ln[N_D-1]}.$$

In large populations of equal size with relatively weak selection, the rate of fixation of beneficial mutations will be less in diploids by an amount approximately equal to h.

In summary, if mutations are rare and tend to fix before the appearance of the next mutation (the rate of fixation is faster than the rate of mutation: $\rho_f \gg \rho_m$), then from equations (3.3) and (3.4) we would expect a diploid population to accumulate dominant beneficial mutations more rapidly and a haploid population to accumulate recessive beneficial mutations more swiftly. If beneficial mutations occur sufficiently often (the rate of fixation is slower than the rate of mutation: $\rho_f \ll \rho_m$), however, multiple alleles will segregate simultaneously. Here, we expect that haploid populations would adapt faster since beneficial alleles spread faster in haploid populations (with $h < 1$) and thus more nesting of advantageous alleles within a lineage can occur. For further analytical comparisons, see [31].

For illustration, let us consider the data of Paquin and Adams [34, 35]. They studied populations of *S. cerevisiae*, where population sizes were about 4.9×10^9 and 4.5×10^9 for the haploid and diploid populations, respectively ([35], Adams pers. comm.). Paquin and Adams followed changes in marker frequencies to determine when positive selective sweeps were passing through the populations. They found that the rate of sweeps per cell per generation was approximately 3.6×10^{-12} for haploid cells and 5.7×10^{-12} for diploid cells. That is, approximately 60% more sweeps were observed per diploid cell. These figures imply that a beneficial sweep occurred on average once every 57 generations within the haploid populations and once every 39 generations within the diploid populations.

At first glance, this appears to be a significant advantage. As noted by Paquin and Adams and by Charlesworth [5], however, this advantage would be partially offset by the fact that haploid populations tend to have higher densities, which could, in nature, translate to larger haploid population sizes. The haploid populations in the chemostat experiments of Paquin and Adams were 40% more dense than the diploid populations. If the haploid population had been 40% larger than the diploid population, the measured advantage to the diploid population would have been reduced, presumably, from 60% to about 14%.

Another source of uncertainty in the interpretation of the experimental results is the quantity L, the number of loci subject to beneficial mutations. It need not be the case that the same number of loci are subject to beneficial mutations in haploid and diploid populations, even when the haploids and the diploids are isogenic since they are not isomorphic. Diploid cells, for instance, tend to

		$\sigma = 0.09$	$\sigma = 0.10$
Stochastic:	$t_f(1/N)$	467	422
Deterministic:	$\bar{t}_f(1/N, 1 - 1/N)$	519	469
Deterministic:	$\bar{t}_f(1/N, 0.05)$	215	194

TABLE 3. Time until fixation in generations. Parameters were chosen from the experiments of Paquin and Adams: $N = 5 \times 10^9$, $\sigma = 0.09$ for diploid mutants, and $\sigma = 0.10$ for haploid mutants. For comparison, in the last row, the times required for a mutation to reach a frequency of 5% are given.

be larger with smaller surface area/volume ratios and tend to be less efficient at nutrient uptake [1, 42]. Thus, observing more adaptive sweeps in a diploid population may simply mean that the diploid population was less well adapted to begin with.

Paquin and Adams also estimated the fitnesses of the new beneficial alleles, finding on average that $s = 0.1$ and $h = 0.9$. In Table 3, we apply equations (3.5) and (3.6) to find the expected times to fixation for these mutations ([12], p. 61, presents a similar analysis). Clearly, beneficial mutations with the selective coefficients measured take several hundreds of generations to fix and spend a large portion of this time rare. These fixation times are much too large to correspond to the events (the sweeps) measured by Paquin and Adams, which occurred roughly every 50 generations. Therefore, multiple mutations must have been segregating simultaneously, which is consistent with later interpretations that the populations were polymorphic [2]. If, however, the populations were subject to such frequent mutations, then we might expect the populations with the faster rate of spread of beneficial alleles (the *haploid* populations) to adapt faster (compare equations (3.8) and (3.9)), in direct contrast to the conclusions in [34].

A computer program (available upon request) was developed to refine these arguments. Again for illustration, the parameters of Paquin and Adams were employed. In the simulations, one locus is a marker locus subject to mutations at a rate of 10^{-7}. The remainder of the genome is summarized by the number, n, of beneficial mutations carried relative to the current wildtype, a number which could vary from zero (in the current wildtype) up to eight. Selection favors individuals according to the number of beneficial mutations, with their relative fitness equalling $(1 + \sigma)^n$. Note that only the number of mutations matters and not their location, so that individuals within the same class may have different genotypes but are equally fit. When beneficial mutations are rare, they occur stochastically upon each fitness class according to the frequency of that class. That is, if a number chosen randomly from a uniform distribution between zero and one is less than the product of the mutation rate, the class frequency and the population size, then a mutation occurs moving one individual (with frequency of $1/N$) from the given class (i) to the next highest fitness class ($i + 1$). When beneficial mutations are common (more than one mutation expected in the population for a given fitness class), they occur deterministically. When the current wildtype reaches a low frequency (0.0001), it is grouped with the class carrying one beneficial mutation and then all classes (i) are moved down to the preceding class ($i - 1$). The number of times that the classes are reset is an estimate of the number of beneficial sweeps that have passed through the population. Technically, these sweeps are not of particular alleles but of fitness classes. Thus we are measuring the rate at which better fitness classes replace worse ones, which is a reasonable measure of the rate of adaptation. A less direct estimate of the number of sweeps comes from the changing frequency of the marker allele [34]. The marker is initially absent from the population and increases in frequency according to its mutation rate. When beneficial mutations occur, they tend not to occur on marker chromosomes, which are fairly rare, but rather on non-marker chromosomes. Thus as beneficial sweeps occur, the non-marker chromosomes are at a selective advantage and tend to displace the marker chromosomes from the population. Eventually mutations to the marker begin to accumulate on the lineage with the new beneficial mutation and the marker once again rises in frequency. In keeping with the definition used in the experimental system [34], a sweep was identified when the marker frequency decreased over at least two censuses, where a census occurred approximately once every seven generations. The number of sweeps over a period of 10,000 generations was then determined. For comparison, Paquin and Adams observed 64 changes in marker frequency in 2,612 generations. Taking into account their observation that 1.6 times as many sweeps occurred in the diploid population, this observation translates into 188 sweeps (S_H, say) per 10,000 generations for the haploid populations and 302 sweeps (S_D) per 10,000 generations for the diploid populations.

To evaluate the importance of selection versus mutation on the rate of adaptation in the experiments of Paquin and Adams, simulations were performed using the data previously cited ($N_H = N_D = 5 \times 10^9$, $s = 0.1$, and $h = 0.9$) and allowing the remaining parameter, the rate of beneficial mutations, to vary. From equations (3.3) and (3.4), the diploid mutation rate should be $2h$ times as large as the haploid mutation rate, which allows for both the fact that twice as many mutations would appear in the diploid population and also the fact that proportionately fewer mutations survive loss when they first appear in diploids (since their fitness is smaller by an amount h). The haploid mutation rate was thus varied from 10^{-14} to 10^{-3} per individual ($= 2\mu Ls$ in terms of equation (3.3)); the diploid mutation rate was set equal to $2h$ times this quantity.

Only with the very smallest mutation rates ($\leq 10^{-13}$) did more mutations accumulate in the diploid population by an amount approaching $2h$. With such low mutation rates, mutations did not tend to cosegregate but rather accumulated independently. Note that from our analysis, we expect this independence when

the rate of fixation is faster than the rate of mutation; in accord with these simulations, $\rho_f \approx 0.002 \gg \rho_m$ implies that $2\mu L s \ll 4 \times 10^{-13}$. When the mutation
rate is this low, however, very few sweeps (< 10) occurred in 10,000 generations.
Clearly this range of mutation rates is inappropriate for the observed number of
sweeps ($S_H = 188$ and $S_D = 302$).

The number of sweeps observed for higher mutation rates is shown in Figures
6 and 7. In Figure 6, the number of times that the fitness classes were reset is shown, while, in Figure 7, the number of sweeps estimated by tracking the
marker frequency is shown. For very high mutation rates (higher than 10^{-6}), the
marker frequency was a very poor indicator of the number of beneficial sweeps;
the marker frequency tended to increase and decrease very rapidly, with these
fluctuations becoming uncoupled from the increase of any particular beneficial
mutation. Over the entire range of parameters illustrated, haploids and diploids
performed remarkably similarly. That is, even though diploid individuals experienced a mutation rate that was 80% larger than that of haploids, the fact that
fitness was higher in haploids by only 1% was enough to compensate for this
mutation difference. In short, in the range examined, small selective differences
were enough to balance large differences in mutation rates, such that diploids
did not adapt at a faster rate in the simulations.

Notice that, in Figure 7, the maximum number of sweeps observed in both
the haploid and diploid populations occurred at mutation rates near 10^{-6}. This
maximum corresponded to only 150 sweeps per 10,000 generations (as measured
by changes in marker frequency), which is below the observed number of sweeps
for both populations and far below the observed number for the diploid populations (again, $S_H = 188$ and $S_D = 302$). Nevertheless, taking 10^{-6} as the
best estimate for the genome-wide rate of beneficial mutations, the simulations
confirm that under the experimental conditions diploid populations would not
have adapted faster than haploid populations simply because there were twice as
many genes that could mutate beneficially. Further experiments are needed to
shed light on this stark contradiction between the theoretical and experimental
results and to determine, in other organisms, the rate of beneficial mutations
and their selective consequences.

3.2.2. *Interbreeding haploid and diploid populations.* We now examine sweeps
of beneficial mutations occurring within a sexual population and determine their
influence on ploidy levels. We return to the life cycle illustrated in Figure 1,
with genes at the ploidy locus altering the proportion of the life cycle spent
in the haploid versus diploid phase. We now consider the impact of a single
new beneficial mutation with selective coefficients given in Table 2. Let us first
calculate the mean fitness among haploids and diploids. In any generation, if the
mutation is at a frequency of p *among all genotypes at the ploidy locus*, then the
mean fitness among individuals experiencing selection entirely as diploids would

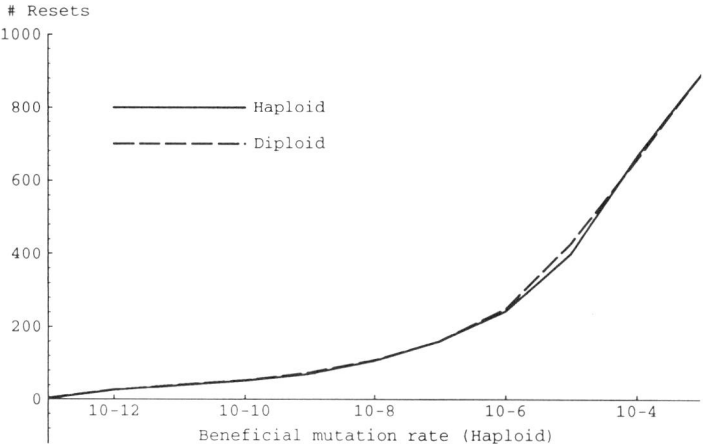

FIGURE 6. The number of sweeps observed in 10,000 generations, as evidenced by the disappearance of the wildtype class and the resetting of fitness classes. The selective advantage of mutations was 0.09 for diploids and 0.1 for haploids. The mutation rate appropriate to haploid individuals is given along the abscissa; the diploid mutation rate was obtained by multiplying this quantity by $2h$ (1.8).

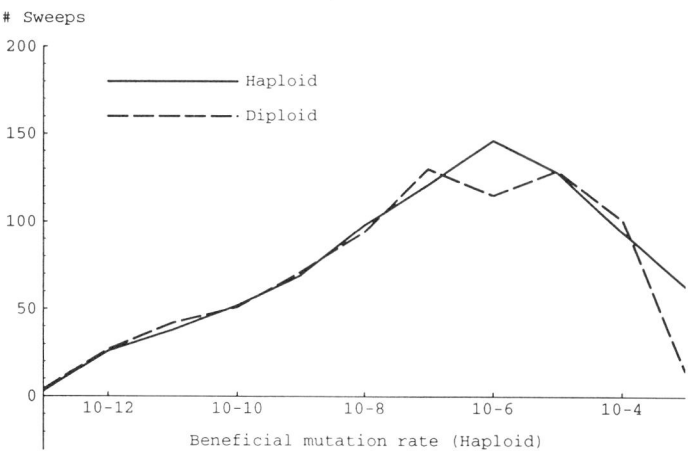

FIGURE 7. The number of sweeps observed in 10,000 generations, as evidenced by changes in the marker frequency. See caption for Figure 6.

be

$$\begin{aligned}\bar{w} &= (1+s)p^2 + (1+hs)2p(1-p) + (1-p)^2 \\ &= 1 + sp[p + 2h(1-p)],\end{aligned}$$

(3.10)

compared to

(3.11) $$\bar{w} = (1+s)p + (1-p) = 1 + sp,$$

for individuals experiencing entirely haploid selection. It is easy to show that the mean fitness of diploids is higher than that of haploids when $h > 1/2$; conversely, the mean fitness of haploids is higher when $h < 1/2$. Even though a haploid that carries a beneficial mutation will tend to leave more offspring, a diploid has twice as many chances to carry a mutation; these factors balance at $h = 1/2$ such that haploids and diploids within a single interbreeding population have the same mean fitness (as long as the gene frequency is the same for both). For individuals spending a proportion (t) of the life cycle in the diploid phase and the remaining time $(1 - t)$ in the haploid phase, the mean fitness is

(3.12) $$\bar{w} = (1+s)p^2 + (1+hs)^t[(1+s)^{(1-t)} + 1]p(1-p) + (1-p)^2.$$

For weak selection (s near zero), the mean fitness is an increasing function of the amount of time spent in the diploid phase for dominant beneficial mutations ($h > 1/2$) but a decreasing function of t for recessive mutations ($h < 1/2$). For strong selection, the mean fitness is a more complex function of the parameters, but as a first approximation we expect the more diploid members of a population to have a higher average fitness and to increase in frequency during the sweep of new *dominant* beneficial mutations, while we expect the more haploid members of a population to increase in frequency during the sweep of new *recessive* mutations. These arguments, however, ignore the development of genetic associations, which may have a large impact on the evolution of ploidy levels, as was the case for deleterious mutations. It need not be the case that during the evolutionary process, the frequency of beneficial mutations is the same in the more haploid individuals and the more diploid individuals. Through simulations, we follow changes in ploidy levels that occur in response to sweeps of beneficial mutations. We find that genetic associations do develop that influence life cycle evolution.

The population evolves according to the recursions given in the Appendix A assuming positive selection and no recurrent mutations ($\mu_1 = \mu_2 = 0$). Initially, we will assume that the population is fixed upon a wildtype viability allele (A). In the absence of variation at the viability locus, evolution at the ploidy locus is neutral so that any combination of ploidy alleles is neutrally stable. The population under consideration thus begins with some positive frequency of both the C_1 and C_2 ploidy alleles ($0 < x_1, x_3 < 1; x_1 + x_3 = 1$). We then introduce a small amount of the beneficial mutation (a) and iterate the recursions until this new allele is near fixation within the population. At this point, we determine how the ploidy alleles have changed in frequency. A particular ploidy level is said to be *favored*, when alleles that cause individuals to spend more time within this phase increase in frequency. Directional ploidy determination ($d_{11} < d_{12} < d_{22}$ or $d_{11} > d_{12} > d_{22}$) was assumed to simplify the interpretation of the results. We start by presenting results for the case when the new beneficial mutation is begun on both the C_1 and C_2 chromosomes according to their frequencies within the population (no initial disequilibrium). In this case, while the frequency of mutant *chromosomes* does not differ between haploids and diploids, the frequency of mutant *individuals* is higher in diploids, since diploid individuals have two alleles either of which may be mutants. We then present and discuss results for starting conditions with linkage disequilibrium. Disequilibrium would be expected if mutations tend to occur only once, initially with either the C_1 or the C_2 allele.

For any given rate of recombination and strength of selection, a value for the degree of dominance of the beneficial mutation (h) could be found above which increases in the diploid phase were favored and below which increases in the haploid phase were favored. This phenomenon is illustrated in Figure 8 for a particular starting position ($x_1 = 0.499$, $x_2 = 0.001$, $x_3 = 0.499$, $x_4 = 0.001$) and a particular set of ploidy alleles ($d_{11} = 0.499$, $d_{12} = 0.500$, $d_{22} = 0.501$ as in Figure 2). Twelve different combinations of ploidy alleles and starting positions (chosen to be different; all with no linkage disequilibrium) were similarly analyzed. Each produced slightly different cut-off values, but the differences were so slight as to make no qualitative and few quantitative differences to the graphs shown in Figure 8. These graphical analyses indicate that the evolution of diploidy is more often favored when beneficial mutations are dominant as expected, but that dominance was not a sufficient condition for the evolution of diploidy. In fact, with tight linkage or strong selection, haploidy was often favored despite the fact that the new beneficial mutations were dominant. This dependence on recombination implies that genetic associations do have a role in the evolution of ploidy levels during the spread of beneficial mutations in sexual populations.

Genetic associations were followed from one thousand random starting conditions (d_{ij} randomly chosen so as to be directional; x_i randomly chosen such that there was no linkage disequilibrium and so that $x_1 + x_3 \approx 0.99$) for each of five rates of recombination ($r \in 0, 0.01, 0.1, 0.25, 0.5$). Each run began with no disequilibrium and yet each run soon developed disequilibrium that favored the more haploid ploidy allele (negative disequilibrium if $d_{11} < d_{12} < d_{22}$ and positive disequilibrium if $d_{11} > d_{12} > d_{22}$). The simulations were then repeated with the linkage disequilibrium artificially reset to zero each generation using that generation's gene frequencies. By this method, we could remove the influence of disequilibrium on ploidy evolution for comparison. Under these conditions and for all rates of recombination, diploidy was favored when the beneficial mutations were dominant, while haploidy was favored when the beneficial mutations were recessive (the exact cut-off was slightly different from $h = 1/2$, but was consistent

S. P. OTTO

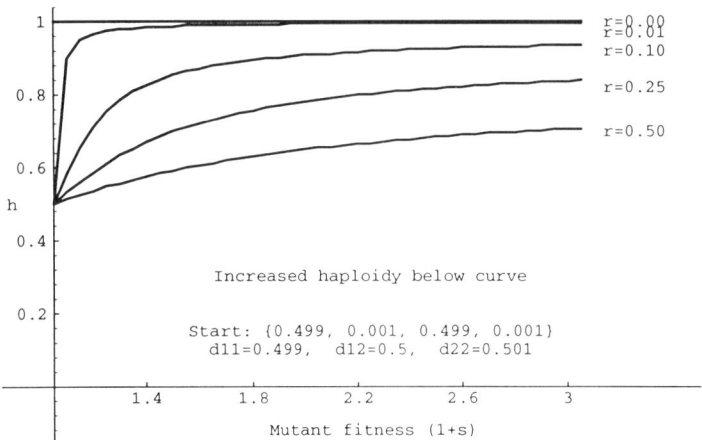

FIGURE 8. Parameters favoring the evolution of increased haploidy (below curve) or increased diploidy (above curve) for various rates of recombination (no initial disequilibrium). Nearly identical curves were produced with different starting positions (in linkage equilibrium) or with different ploidy alleles.

with equation (3.12)). The evolution of ploidy levels in response to beneficial mutations thus exhibits very similar behavior to the evolution of ploidy levels in response to deleterious mutations: ignoring disequilibrium, masking deleterious recessive mutations or revealing dominant beneficial mutations favors diploidy; genetic associations develop, however, that tend to favor the evolution of haploidy especially when linkage is tight.

A single sweep of a beneficial mutation can have a dramatic effect on the frequencies of the ploidy alleles within a population. In Table 4, the change in the frequency of a haploid ploidy allele is given for a range of parameter values ($h = 1/2$ so that haploidy is expected to be favored). A fairly rare haploid ploidy allele can quadruple in frequency, while more common ploidy alleles can traverse halfway across the frequency space towards fixation. These dramatic changes were observed when the rate of recombination was low and selection was weak. With low rates of recombination, weak selection is more effective in changing the frequencies of the ploidy alleles, while the opposite is true with high rates of recombination. The extent of frequency change appears to be related to the strength of genetic associations that are created and to the number of generations that these associations are maintained.

Heretofore, we have considered beneficial mutations that arise in a population in linkage equilibrium. The evolution of ploidy levels in response to sweeps of beneficial mutations is, however, sensitive to the initial level of linkage disequilibrium. With $d_{11} < d_{12} < d_{22}$, if the linkage disequilibrium is initially negative, then the beneficial mutation arises more frequently than expected on the haploid

		Final freq(C_1) when $1 + s$ equals:			
	Initial freq(C_1)	1.20	1.50	2.00	10.0
$r = 0.0$	0.500	0.782	0.768	0.750	0.667
$r = 0.5$	0.500	0.509	0.520	0.532	0.566
$r = 0.0$	0.050	0.200	0.181	0.162	0.103
$r = 0.5$	0.050	0.052	0.054	0.057	0.065
$r = 0.0$	0.950	0.983	0.982	0.981	0.973
$r = 0.5$	0.950	0.952	0.954	0.956	0.961

TABLE 4. Extent of allele frequency change. The table shows the frequency of the C_1 allele (the more haploid allele) after a single sweep for three different starting frequencies of the C_1 allele (0.5, 0.05, and 0.95), two recombination rates (0 and 0.5), and four fitness values ($1 + s = 1.20$, $1 + s = 1.50$, $1 + s = 2.00$, $1 + s = 10.00$). The remaining parameters were $d_{11} = 0.2$, $d_{12} = 0.5$, $d_{22} = 0.8$, and $h = 0.5$. Under these conditions, haploidy is always favored as illustrated in Figure 8, but the extent of change depends on s and r.

allele (C_1) and haploidy is favored in a larger area of the parameter space. With positive disequilibrium, diploidy is favored for more combinations of parameters. In Figures 9 and 10, the curves delimiting the parameter space in which haploidy is favored from the space in which diploidy is favored are given for rather extreme initial disequilibrium values (only three of the four possible haplotypes exist within the starting population). When $r = 0$, the ploidy allele on which all beneficial mutations occur will sweep to fixation, regardless of its ploidy level. With higher rates of recombination, the selective regime will matter as well as the initial population composition. These considerations lead us to speculate that ploidy levels may fluctuate more often in populations with tight linkage (or other mechanisms that maintain disequilibria), since these fluctuations will simply reflect the random appearance of beneficial mutations on chromosomes with different ploidy alleles. With loose linkage or when beneficial mutations tend to appear in linkage equilibrium, the above results lead us to expect the evolution of haploidy when r is small, h is small, *or* s is large and to expect the evolution of diploidy when r is large, h is large, *and* s is small.

We clearly need more experimental evidence on the rate of beneficial mutations and their selective consequences. Only with this information can we determine whether sweeps of mutations tend to favor the evolution of diploidy in a particular organism. We know now, however, that beneficial mutations do not unequivocally favor the evolution of diploidy in sexual populations. Even though diploids have twice the opportunity to carry new mutations (as heterozygotes), the fact that these mutations can rise in frequency faster in haploids (especially when h is low) can overwhelm this advantage.

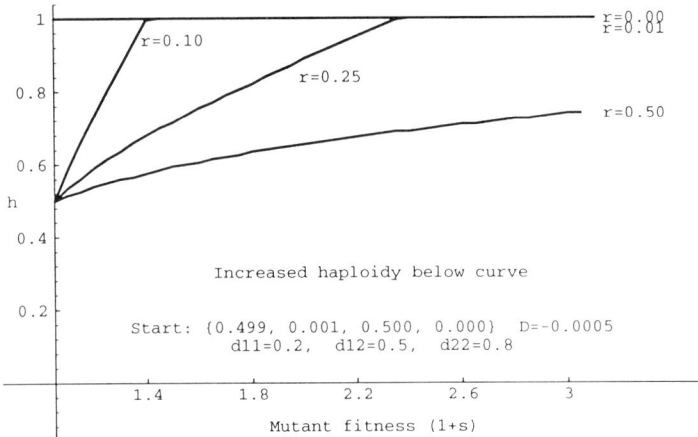

FIGURE 9. Parameters favoring the evolution of haploidy (below curve) or diploidy (above curve) for various recombination rates ($D = -0.0005$). Here $d_{11} = 0.2$, $d_{12} = 0.5$, and $d_{22} = 0.8$. If the d_{ij}s are chosen as in Figure 8 (closer together) then haploidy is favored even more often (haploidy favored for all h and r when $1 + s > 1.25$).

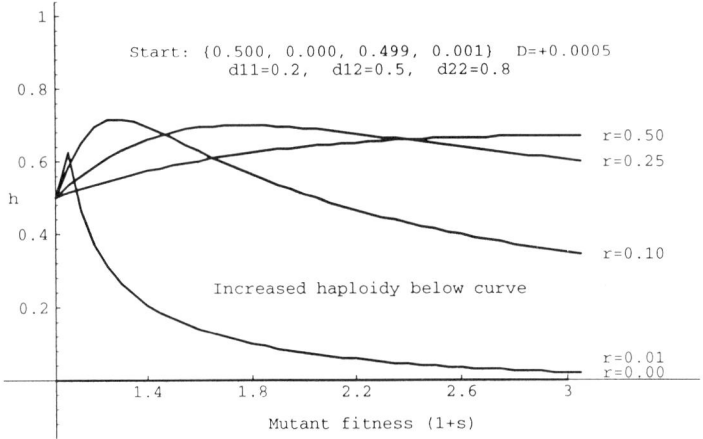

FIGURE 10. Parameters favoring the evolution of haploidy (below curve) or diploidy (above curve) for various recombination rates ($D = +0.0005$). Here $d_{11} = 0.2$, $d_{12} = 0.5$, and $d_{22} = 0.8$. If the d_{ij}s are chosen as in Figure 8 (closer together) then diploidy is favored even more often (diploidy favored for all h and r when $1 + s > 2.35$).

4. Conclusions

In a series of papers published during the 1920's [**17, 18**], Haldane showed that selection would occur most rapidly in populations that are either asexual or haploid. Having examined various forms of selection in amphimictic diploids (including autosomal, sex-linked, sex-limited and familial selection), Haldane concluded that "selection proceeds more slowly with all other systems of inheritance" [**17**]. The fact that allele frequencies change faster in response to selection acting on haploid rather than diploid genotypes in sexual organisms has not, until recently, been incorporated into theories about life cycle evolution. We now proceed to reiterate and interpret the main results in light of this fact.

It has generally been assumed that life cycles evolve as if allele frequencies were changing in a uniform manner throughout a population. Assuming that allele frequencies are equivalent among more haploid and more diploid members of a sexual population, equations (3.10) and (3.11) show that the mean fitness of diploids is higher than the mean fitness of haploids when deleterious mutations are recessive (masked) and when beneficial mutations are dominant (revealed). This leads one to expect that masking deleterious mutations and revealing beneficial mutations should favor the evolution of diploidy. This conclusion is erroneous, however, since allele frequencies will not remain uniform throughout a mixed population of haploids and diploids but will develop patterns that reflect the argument of Haldane. Specifically, allele frequencies will respond faster to selection among individuals which experience longer periods of haploid selection, with deleterious alleles decreasing in frequency and beneficial alleles increasing in frequency at faster rates. Put another way, since haploid individuals do not mask their mutations from selection, individuals that survive a long period of haploid selection tend to bear fewer offspring carrying deleterious alleles and more offspring carrying beneficial alleles than individuals that survive diploid selection. In other words, genetic associations tend to develop that couple alleles which increase the haploid phase with viability alleles that are the most fit.

Mechanisms that maintain genetic associations within a sexual population favor the expansion of the haploid phase in the life cycle. The most obvious such mechanism is tight linkage, but assortative mating, selfing, and asexual reproduction in populations with life cycle variation have similar influences on the evolution of ploidy levels [**33**]. For the increase of the diploid phase of a life cycle to be favored in an organism, mating must be fairly random within the population, recombination rates must be sufficiently high, and mutations must tend to be recessive when deleterious or dominant when beneficial.

Haldane's results also provide insight when we compare separate haploid and diploid asexual populations. Now selection is less effective in diploids not because of amphimixis, but because the effects of beneficial mutations are somewhat diluted by the presence of a second homologous allele within the genome (assuming $h < 1$). Thus while diploids have twice as many opportunities to bear new beneficial mutations, these mutations will sweep to fixation at slower rates among diploid asexuals. The faster a beneficial mutation sweeps through an asexual population, the sooner new beneficial mutations can accumulate on lineages with previous beneficial mutations. Thus beneficial mutations can be nested within a lineage at a faster rate in haploid asexuals than diploid asexuals. This effect will be unimportant if mutations are rare and fix before the appearance of new mutations but will be very important if beneficial mutations occur often within a population. Even if mutations are rare, haploid populations will still accumulate beneficial mutations more rapidly than diploid populations if these mutations are recessive. This follows from other work by Haldane [19], which indicates that a beneficial mutation which arises in a single individual within a population will survive the initial period of random sampling with a probability that is approximately equal to the selective advantage of the new allele. This probability is $2s$ for new alleles within the haploid population but only $2hs$ in the diploid population. Twice as many mutations will still arise in a diploid population, but, if $h < 1/2$, fewer mutations will survive the first few generations of sampling when compared to a haploid population. The analyses in this paper allow us to evaluate whether mean fitness will increase at a higher rate in asexual haploid populations or in asexual diploid populations. Experiments are needed to determine whether haploid populations are favored under the conditions predicted (e.g. frequent mutations within the population) and are not favored when predicted (e.g. mutations rare within the population). Such experiments could manipulate the mutation rate within a population by either changing the population size or by exposing individuals to mutagens. Experiments are also needed to determine whether mean fitness differences measured in separate populations translate into competitive differences when the haploids and diploids are placed within a single environment.

The conditions outlined in this paper and elsewhere [32, 33] allow us to test the importance of masking deleterious mutations and revealing beneficial mutations to the evolution of ploidy levels. Given information about the life history of an organism and about selection, we can now predict whether the haploid or the diploid phase of the life cycle is likely to dominate. Whether these theories have any explanatory power has yet to be determined by careful comparative studies and experimental work.

5. Acknowledgements

The author wishes to thank Michael Cummings and Allen Orr for many fruitful discussions as well as Cheryl Jenkins, Mark Kirkpatrick, Maria Orive, and Monty Slatkin for innumerable suggestions that improved the presentation of this work.

Appendix A. Recursions.

In this appendix, we present and analyze recursions that monitor changes in the timing of meiosis in sexual organisms that experience both diploid and haploid selection (a regular alternation of generations). We census at the gamete stage of the life cycle, and then follow the four chromosomes (AC_1, aC_1, AC_2, aC_2) through the life cycle, tracking their frequencies over time (x_1, x_2, x_3, x_4, respectively). The recursions that map the population from one generation to the next and which correspond to the model described in the text and in Figure 1 are:

$$
\begin{aligned}
Tx_1' = & \ (1 - \frac{\mu_2}{2})(1 - \frac{\mu_1}{2})[x_1x_1 + (1 \pm hs)^{d_{11}}x_1x_2 + x_1x_3 \\
& + (1 \pm hs)^{d_{12}}(1 - r)x_1x_4 + (1 \pm hs)^{d_{12}}rx_2x_3]
\end{aligned}
$$

$$
\begin{aligned}
Tx_2' = & \ (\frac{\mu_2}{2})(1 - \frac{\mu_1}{2})[x_1x_1 + (1 \pm hs)^{d_{11}}x_1x_2 + x_1x_3 \\
& + (1 \pm hs)^{d_{12}}(1 - r)x_1x_4 + (1 \pm hs)^{d_{12}}rx_2x_3] \\
& + (\frac{\mu_1}{2})[(1 \pm s)^{(1-d_{11})}x_1x_1 + (1 \pm hs)^{d_{11}}(1 \pm s)^{(1-d_{11})}x_1x_2 \\
& \quad + (1 \pm s)^{(1-d_{12})}x_1x_3 + (1 \pm hs)^{d_{12}}(1 \pm s)^{(1-d_{12})}(1 - r)x_1x_4 \\
& \quad + (1 \pm hs)^{d_{12}}(1 \pm s)^{(1-d_{12})}rx_2x_3] \\
& + [(1 \pm hs)^{d_{11}}(1 \pm s)^{(1-d_{11})}x_1x_2 + (1 \pm hs)^{d_{12}}(1 \pm s)^{(1-d_{12})}rx_1x_4 \\
& \quad + (1 \pm s)x_2x_2 + (1 \pm hs)^{d_{12}}(1 \pm s)^{(1-d_{12})}(1 - r)x_2x_3 \\
& \quad + (1 \pm s)x_2x_4]
\end{aligned}
$$

$$
\begin{aligned}
Tx_3' = & \ (1 - \frac{\mu_2}{2})(1 - \frac{\mu_1}{2})[x_1x_3 + (1 \pm hs)^{d_{12}}rx_1x_4 + (1 \pm hs)^{d_{12}}(1 - r)x_2x_3 \\
& + x_3x_3 + (1 \pm hs)^{d_{22}}x_3x_4]
\end{aligned}
$$

$$
\begin{aligned}
Tx_4' = & \ (\frac{\mu_2}{2})(1 - \frac{\mu_1}{2})[x_1x_3 + (1 \pm hs)^{d_{12}}rx_1x_4 + (1 \pm hs)^{d_{12}}(1 - r)x_2x_3 \\
& + x_3x_3 + (1 \pm hs)^{d_{22}}x_3x_4] \\
& + (\frac{\mu_1}{2})[(1 \pm s)^{(1-d_{12})}x_1x_3 + (1 \pm hs)^{d_{12}}(1 \pm s)^{(1-d_{12})}rx_1x_4 \\
& \quad + (1 \pm hs)^{d_{12}}(1 \pm s)^{(1-d_{12})}(1 - r)x_2x_3 + (1 \pm s)^{(1-d_{22})}x_3x_3 \\
& \quad + (1 \pm hs)^{d_{22}}(1 \pm s)^{(1-d_{22})}x_3x_4] \\
& + [(1 \pm hs)^{d_{12}}(1 \pm s)^{(1-d_{12})}(1 - r)x_1x_4 + (1 \pm hs)^{d_{12}}(1 \pm s)^{(1-d_{12})}rx_2x_3 \\
& \quad + (1 \pm s)x_2x_4 + (1 \pm hs)^{d_{22}}(1 \pm s)^{(1-d_{22})}x_3x_4 + (1 \pm s)x_4x_4]
\end{aligned}
$$

where T is the sum of the right hand sides. In these recursions, $\mu_1/2$ measures the mutation rate during meiosis (when diploids produce haploid offspring) and $\mu_2/2$ measures the mutation rate during gametogenesis (when haploids produce gametes). Except to evaluate the role played by the position of mutations, we assume that $\mu_1 = \mu_2 = \mu$. The $\pm$ sign should read as a positive sign when beneficial mutations are considered and as a negative sign when deleterious mutations are considered.

Appendix B. Analysis with recurrent deleterious mutations.

The equilibrium frequencies when the C_1 allele is fixed are:

(B.1)
$$\hat{x}_1 = 1 - \hat{x}_2$$
$$\hat{x}_2 = \frac{\mu[1 + (1 - s)^{(1-d_{11})}]}{2[1 - (1 - s)^{(1-d_{11})}(1 - hs)^{(1-d_{11})}]}.$$

Invasion of the C_2 allele from near this equilibrium is governed by the leading eigenvalue obtained from the recursions linearized in the vicinity of the equilibrium (B.1) (see [32]). This leading eigenvalue equals:

(B.2)
$$\lambda_L = 1 - \mu K_1 + r\mu K_2 + O(\mu^2)$$

where

$$K_1 = \frac{[1 + (1 - s)^{(1-d_{11})}][(1 - hs)^{d_{11}} - (1 - hs)^{d_{12}}]}{2[1 - (1 - s)^{(1-d_{11})}(1 - hs)^{d_{11}}]}$$

$$K_2 = \frac{(1 - s)^{(1-d_{11}-d_{12})}(1 - hs)^{d_{12}} K_3}{2[1 - (1 - s)^{(1-d_{11})}(1 - hs)^{d_{11}}][1 - (1 - r)(1 - s)^{(1-d_{12})}(1 - hs)^{d_{12}}]}$$

$$K_3 = (1 - s)^{d_{11}} - (1 - s)^{d_{12}} - (1 - s)(1 - hs)^{d_{11}} - (1 - s)^{d_{12}}(1 - hs)^{d_{11}}$$
$$+ (1 - s)(1 - hs)^{d_{12}} + (1 - s)^{d_{11}}(1 - hs)^{d_{12}}.$$

K_1, K_2, and K_3 all have the same sign as $(d_{12} - d_{11})$. An inspection of the leading eigenvalue indicates that haploidy is favored whenever $r = 0$, but that with sufficiently strong masking and high recombination diploidy is favored (see equation (B.3)).

Quantitatively, the invasion of a new allele increasing the diploid phase of the life cycle will occur if

$$rW_1^{d_{12}}[W_2^{d_{11}} - W_2^{d_{12}} - 2W_2 W_1^{d_{11}} - W_2^{d_{11}} W_1^{d_{11}}$$
$$- W_2^{d_{12}} W_1^{d_{11}} + 2W_2 W_1^{d_{12}} + 2W_2^{d_{11}} W_1^{d_{12}}]$$
(B.3) $$> W_2^{d_{12}}[1 + W_2^{(1-d_{11})}][W_1^{d_{11}} - W_1^{d_{12}}][1 - W_2^{(1-d_{12})} W_1^{d_{12}}],$$

where $W_1 = (1 - hs)$ and $W_2 = (1 - s)$. Conversely, a new allele increasing the haploid phase of the life cycle will invade if this inequality is reversed.

Where in the life cycle mutations occur does not affect the qualitative behavior of the model with respect to linkage. It does, however, affect the quantitative results, especially when s is large. It can be shown that the evolution of diploidy is facilitated when mutations occur during meiosis (sporulation) as opposed to gametogenesis. This effect may be understood by the fact that mutations which occur during sporulation are subjected to haploid selection before the diploids are formed, thus reducing the frequency of mutations carried by the diploids.

REFERENCES

1. ADAMS, J. AND HANSCHE, P. E. 1974. Population studies in microorganisms I. Evolution of diploidy in *Saccharomyces cerevisiae*. Genetics 76: 327–338.

2. ADAMS, J. AND OELLER, P. W. 1986. Structure of evolving populations of *Saccharomyces cerevisiae*: Adaptive changes are frequently associated with sequence alterations involving mobile elements belonging to the Ty family. Proc. Nat. Acad. Sci USA 83: 7124-7127.

3. BELL, G. 1982. The Masterpiece of Nature: The Evolution and Genetics of Sexuality. University of California Press, Berkeley.

4. CAVALIER-SMITH, T. 1978. Nuclear volume control by nucleoskeletal DNA, selection for cell volume and cell growth rate, and the solution of the DNA C-value paradox. J. Cell Sci. 34: 247–278.

5. CHARLESWORTH, B. 1983. Adaptive evolution in the laboratory. Nature 302: 479–480.

6. CROW, J. AND KIMURA, M. 1965. Evolution in sexual and asexual populations. Am Nat. 99: 439–450.

7. D'AMATO, F. 1977. Nuclear Cytology in Relation to Development. Cambridge University Press, Cambridge.

8. DARLINGTON, C. D. 1958. The Evolution of Genetic Systems. Basic Books, New York.

9. DESTOMBE, C., VALERO, M., VERNET, P., AND COUVET, D. 1989. What controls haploid-diploid ratio in the red alga, *Gracilaria verrucosa*? J. Evol. Biol. 2: 317–338.

10. EBERSOLD, W. T. 1967. *Chlamydomonas reinhardi:* Heterozygous diploid strains. Science 157: 447–449.

11. ESHEL, I. AND FELDMAN, M. W. 1982. On evolutionary genetic stability of the sex ratio. Theor. Pop. Biol. 21: 430–439.

12. EWENS, W. J. 1969. Population Genetics. Methuen & Co. LTD, London.

13. EWENS, W. J. 1979. Mathematical Population Genetics. Springer-Verlag, New York.

14. FOWELL, R. R. 1969. Life cycles in yeast. *In* A. H. Rose and J. S. Harrison (Eds.), The Yeasts (pp. 461–471). Academic Press, New York.

15. GOFF, L. J. AND COLEMAN, A. W. 1990. DNA: Microspectrofluorometric studies. *In* K. M. Cole and R. G. Sheath (Eds.), Biology of the Red Algae (pp. 43–71). Cambridge University Press, Cambridge.

16. GOLDSTEIN, D. 1992. Heterozygote advantage and the evolution of a dominant diploid phase. Genetics 132: 1195–1198.

17. HALDANE, J. B. S. 1924. A mathematical theory of natural and artificial selection, Part I. Trans. Camb. Phil. Soc. 23: 19–40.

18. HALDANE, J. B. S. 1926. A mathematical theory of natural and artificial selection, Part III. Proc. Camb. Phil. Soc. 23: 363–372.

19. HALDANE, J. B. S. 1927. A mathematical theory of natural and artificial selection, Part V: Selection and Mutation. Proc. Camb. Phil. Soc. 23: 838–844.

20. HALDANE, J. B. S. 1990. The Causes of Evolution. Princeton University Press, Princeton.

21. HOSHAW, R. W., WANG, J.-C., MCCOURT, R. M., AND HULL, H. M. 1985. Ploidal changes in clonal cultures of *Spirogyra communis* and implications for species definition. Amer. J. of Bot. 72: 1005–1011.

22. HUXLEY, J. 1964. Evolution: The Modern Synthesis. John Wiley & Sons, Inc., New York.

23. JENKINS, C. D. 1993. Selection and the evolution of genetic life cycles. Genetics 133: 401–410.

24. KIMURA, M. AND CROW, J. F. 1969. Natural selection and gene substitutions. Genet. Res. Camb. 13: 127–141.

25. KIMURA, M. AND OHTA, T. 1969. Average number of generations until fixation of a mutant gene in a finite population. Genetics 61: 763–771.

26. KIRKPATRICK, M. AND JENKINS, C. D. 1989. Genetic segregation and the maintenance of sexual reproduction. Nature 339: 300–301.

27. KONDRASHOV, A. AND CROW, J. 1991. Haploidy or diploidy: which is better? Nature 351: 314–315.

28. LOCK, R. H. 1906. Recent progress in the study of variation, heredity, and evolution. E. P. Dutton & Co., New York.

29. MAYR, E. 1982. The Growth of Biological Thought. The Belknap Press of Harvard University Press, Cambridge.

30. MULLER, H. J. 1932. Some genetic aspects of sex. Am. Nat. 66: 118–138.

31. ORR, H. A. AND OTTO, S. 1993. Does diploidy increase the rate of adaptation? Genetics: Submitted.

32. OTTO, S. AND GOLDSTEIN, D. 1992. Recombination and the evolution of diploidy. Genetics 131: 745–751.

33. OTTO, S. AND MARKS, J. 1994. Masking mutations from selection: A boon or a bane? Manuscript 000: 000.

34. PAQUIN, C. AND ADAMS, J. 1983a. Frequency of fixation of adaptive mutations is higher in evolving diploid than haploid yeast populations. Nature 302: 495–500.

35. PAQUIN, C. AND ADAMS, J. 1983b. Relative fitness can decrease in evolving asexual populations of *S. cerevisiae*. Nature 306: 368–371.

36. PERROT, V., RICHERD, S., AND VALERO, M. 1991. Transition from haploidy to diploidy. Nature 351: 315–317.

37. RAPER, J. R. AND FLEXER, A. S. 1970. The road to diploidy with emphasis on a detour. Symp. Soc. Gen. Microbiol. 20: 401–432.

38. SCHMALHAUSEN, I. I. 1949. Factors of Evolution: The Theory of Stabilizing Selection. The Blakiston Co., Philadelphia.

39. SIMMONS, M. J. AND CROW, J. F. 1977. Mutations affecting fitness in *Drosophila* populations. Ann. Rev. of Genetics 11: 49–78.

40. STRASBURGER, E. 1894. The periodic reduction of the number of the chromosomes in the life-history of living organisms. Ann. Bot. 8: 281–316.

41. SVEDELIUS, N. 1926. An evaluation of the structural evidence for genetic relationships in plants: Algae. Proc. Int. Cong. Pl. Sci. Ithaca 1: 457–471.

42. WEISS, R. L., KUKORA, J. R., AND ADAMS, J. 1975. The relationship between enzyme activity, cell geometry, and fitness in *Saccharomyces cerevisiae*. Proc. Nat. Acad. Sci USA 72: 794–798.

43. WIENER, P., FELDMAN, M. W., AND OTTO, S. P. 1992. On genetic segregation and the evolution of sex. Evolution 46: 775–782.

DEPARTMENT OF INTEGRATIVE BIOLOGY, UC BERKELEY, BERKELEY CA 94720
Current address: Department of Integrative Biology, UC Berkeley, Berkeley CA 94720
E-mail address: sarah@lolich.berkeley.edu

Lectures on Mathematics in the Life Sciences
Volume **25**, 1994

Genetic Error, Heterozygosity and the Evolution of the Sexual Life Cycle

Richard E. Michod and Todd Gayley

ABSTRACT. Genetic error is a fundamental problem with which all living systems must cope. The two kinds of genetic error, mutation and damage, have different effects in the cell and may play different roles in the evolution of the sexual life cycle. In simple systems, the need to cope with DNA damage selects for diploidy while the need to replicate efficiently selects for haploidy. The models discussed here show that under a range of conditions it is advantageous to alternate between the diploid and haploid stages as is done in the sexual cycle. The advantage of haploid replication became less important as the diploid phase of the sexual cycle became dominant. Once diploidy emerged as the dominant stage, deleterious recessive mutations masked in the diploid stage accumulate. Overdominance at individual loci is also possible. Both of these factors make heterozygosity at individual loci advantageous and this advantage may serve to maintain the outcrossing aspects of the sexual cycle, assuming that recombination is maintained for the function of DNA repair. We develop heuristics for understanding recent results of modifier models of mating system evolution that bear on these ideas.

1. Background

There are two kinds of genetic error: mutation and damage. Both kinds of error involve changes to the DNA molecule, however they differ in their basic nature and their consequences in evolution. A mutation is a change in the base pair sequence of a DNA molecule; however, the changed DNA molecule is still a sequence of standard base pairs. Examples of mutations include deletions, additions, substitutions and rearrangements of the standard base pairs. A damage is a change in the structure of the DNA molecule so that the changed molecule is no longer a regular sequence of the standard base pairs. Examples include breaks, crosslinks, dimers, loss of bases and changes to the chemical compositions of the bases. Because a mutation is a sequence of standard base pairs, it is replicated but cannot be recognized or repaired directly at the level of the DNA molecule (once the mutation is present in both DNA strands). Only natural selection may increase or decrease the frequency of mutations in populations according to their effects on fitness. In addition, mutations may

1991 Mathematics Subject Classification. 92D10, 92D15, 92D25, 92B05

Research supported by grant HD 19949 from the NIH. The authors appreciate the comments and criticisms of Drs. D. Charlesworth, D. Hall, K. Holsinger, M. Kirkpatrick, S. Otto, and M. Uyenoyama.

This paper is in final form and no version of it will be submitted for publication elsewhere.

accumulate in a population, if they are recessive or if there effects on fitness are synergistic (Kondroshov, this volume). In contrast to mutation, a damaged DNA molecule contains "letters" that are not part of the standard DNA alphabet. Consequently, damages cannot be replicated, but they may be recognized and repaired directly in the DNA molecule, so long as there is a backup copy of the information available. Single-strand damages may be repaired by excision repair, whereby the damage is excised and information copied from the good strand. For double-strand damages to be repaired the cell must be diploid, at least for the damaged locus, and recombination must occur to replace the damaged information using good information from the undamaged chromosome. Damages may give rise to mutations if they go unrepaired. Damages do not accumulate in a population, since they usually interfere with replication and transcription.

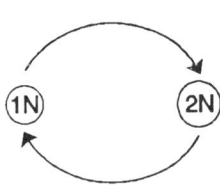

Figure 1. Sexual life cycle ---- the alternation between haploid and diploid cell stages.

There are three aspects of sexual life cycles that we consider: recombination between DNA molecules, the alternation of haploid and diploid cell states and the system of mating that creates diploid cells from haploid cells. We use recombination in its modern molecular sense as the breakage and rejoining of DNA molecules. In eukaryotes, sexual life cycles involve an alternation between haploid and diploid cell stages (Figure 1). A diversity of sexual life cycles exists according to the time spent growing vegitatively in the diploid and haploid cell stage (Bell, this volume). Sexual haploids (like yeast) spend most of their time in the haploid stage, while the diploid stage is transient. In contrast, sexual diploids (like humans) spend most time in the diploid stage, while the haploid stage, the gamete stage, is transient. Regardless of the time spent in each state, all sexual life cycles involve an alternation between haploid and diploid cell states. Sexual life cycles also exist in prokaryotes. For example, simultaneous infection of a cell by multiple phage leads to diploidy, sometimes polyploidy, for all phage genes. In the case of bacterial transformation, the diploid stage occurs for just a few loci at a time, as fragments of homologous DNA are brought into the cell during competence.

Although our ultimate goal is to understand the selective factors molding the diversity of sexual life cycles, our interest here is in the sexual life cycle itself, why it evolved in simple systems and why it is maintained in more complex organisms. Although we think it unlikely that just a few factors could explain all aspects of sexual life cycles, we focus on genetic error and heterozygosity.

2. Recombination

Repair of genetic damage has been argued to be the primary function of molecular recombination (for review, see Bernstein, Hopf and Michod 1987). The kinds of data that support this claim are given in Figure 2. Organisms deficient in homologous recombination are extremely sensitive to DNA damaging agents. Cells or organisms are more sensitive to damaging agents when they are haploid (no genetic redundancy) than when they are diploid. DNA damaging agents increase levels of recombination in all organisms studied. If non-recombinational pathways of repair are impaired, there is a further increase in the response of recombination to DNA damaging agents. In bacteria the recombination protein has two functions: it regulates

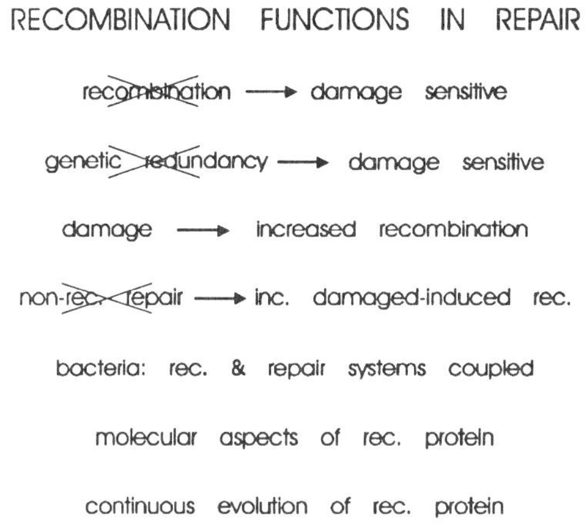

RECOMBINATION FUNCTIONS IN REPAIR

recombination ⟶ damage sensitive

genetic redundancy ⟶ damage sensitive

damage ⟶ increased recombination

non-rec.<repair ⟶ inc. damaged-induced rec.

bacteria: rec. & repair systems coupled

molecular aspects of rec. protein

continuous evolution of rec. protein

Figure 2. Summary of Data that Recombination Functions in DNA Repair. See text for explanation, "rec." = recombination.

the DNA repair capacity of the cell and promotes strand exchange between homologous DNA duplexes. Consequently, the repair and recombination systems are intimately coupled in bacteria. Molecular models of recombination during meiosis are hard to reconcile with the view that recombination evolved to reduce linkage disequilibrium (the traditional view in evolutionary biology) but fit nicely with the DNA repair hypothesis (Bernstein, Hopf and Michod 1988).

There appears to have been a continuous evolution of the recombination protein. The *Escherichia coli* RecA protein functions in homologous genetic recombination and DNA repair and has been extensively studied both *in vivo* and *in vitro* (for review from an evolutionary perspective see Miller and Kokjohn 1990 and Cox 1991). Cox (1991, p. 1296) states his thesis "... is that RecA is a repair system first and a recombination system second." The basis for his conclusion is that "If recombination (genetic diversity) is the primary mission of this system and represents the selective pressure for its evolution and maintenance, then many of the key properties of the RecA system observed in vitro, in particular the filamentous structure and a large investment in ATP hydrolysis, become hard to rationalize...(p. 1296, Cox 1991)."

The primary role of recombination in DNA repair in *E. coli* might be a curiosity, confined to a single bacterium, except for the fact that the RecA protein is conserved in evolution both within bacteria and in eukaryotes. Bacterial RecA proteins have 56 to 100% DNA sequence identity (Roca and Cox, 1990). Although the sequence identity of bacterial RecA to the recombination proteins in bacteriophage T4 and the yeast *Saccharomyces cerevisiae* is lower, around 30%, there is high structural homology of all RecA proteins studied (Story et al. 1993, Ogawa et al. 1993). Story et al. (1993) conclude that "proteins in this group are members of a single family that diverged from a common ancestor that existed prior to the divergence of prokaryotes and eukaryotes." Thus, there appears to have been a continuous evolutionary history of the recombination aspects of sexual life cycles.

The traditional view on the evolution of recombination is that it produces genetic variation by reducing linkage disequilibrium. However, the benefits of reducing linkage disequilibrium are likely to be small for individual organisms, since linkage disequilibrium levels are typically small in natural populations. More importantly, molecular recombination during meiosis usually has no effect on the linkage relationship of genes (for molecular details see Figure 3).

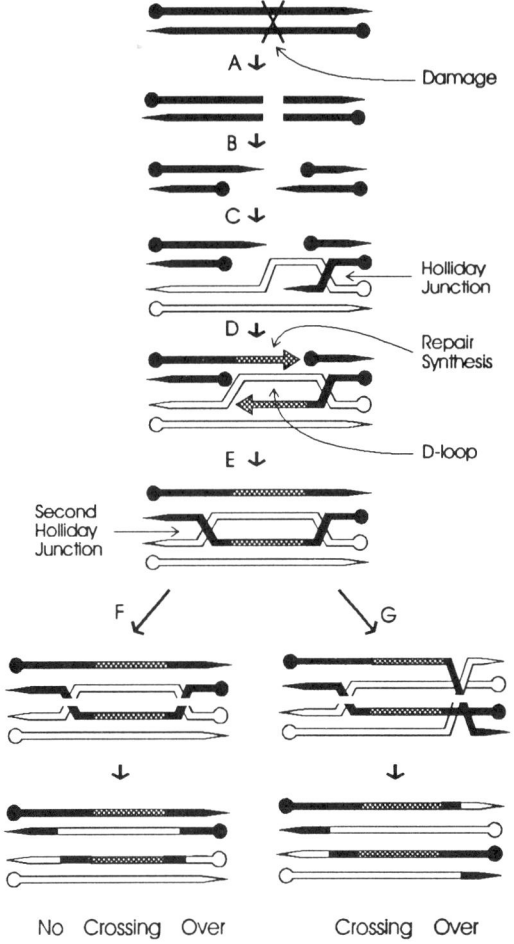

Figure 3. **Double-Strand Break Repair Model of Recombination.** (figure adopted from Bernstein Hopf and Michod (1988) and based on model of Szostak et al. (1983)).

According to the double-strand break repair model (Figure 3), molecular recombination begins with a double strand gap, presumably at the site of a pre-existing damage (Figure 3; step A) Significantly, the initiating molecule with the double strand gap receives information from the other DNA molecule. After strand invasion (Figure 3; Step C) a Holliday junction or "D loop" is formed. A second Holliday junction is formed (Figure 3; Step E). Resolution of two Holliday structures occurs in steps F and G to yield either crossing-over or no crossing-over for loci that flank the site of recombination. A simpler molecular model of recombination is given in Bernstein, Hopf and Michod (1998) that produces crossing-over 100% of the time. This simpler model avoids the second Holliday junction. However, the more complex model shown in Figure 3 is needed to explain the extensive data on gene conversion in fungi.

The fungi data shows that crossing-over occurs between 30-50% of the time during meiosis. When mitotic and meiotic recombination events are considered, the fraction of recombination events that produce new combinations of alleles at linked loci is extremely small (Bernstein Hopf, and Michod 1988). Consequently, the majority of molecular recombination events are cryptic from the point of view of linkage disequilibrium and likely serve another function--we believe that function is DNA repair.

3. Origin of Diploidy and the Sexual Life Cycle

The need for damage repair can explain the origin of diploid cells from haploid cells. The conflicting needs of damage repair and efficient replication can explain the origin of sex in competition with diploidy. These conclusions are supported by the models of Long and Michod (1994) and Michod and Long (1994), that are described in this Section. The original papers should be consulted for a detailed discussion of the assumptions and analysis.

3.1. Consequences of haploidy and diploidy

There are a variety of consequences of haploidy and diploidy that may lead to costs and benefits in terms of fitness. We take the view that an alternation between haploidy and diploidy may be favored so as to reap the benefits of both cell states while reducing their costs.

Diploid cells have two copies of each gene. This genetic redundancy allows for recombinational repair of DNA damage. In addition, alleles at a diploid locus may either be heterozygous or homozygous. Heterozygosity is of central importance to the continued evolution of the sexual cycle according to the ideas discussed in Section 4. Heterozygosity could have played a role in the origin of the sexual cycle, but we believe that damage repair was the primary function of diploid stage of the sexual cycle initially.

Recombinational repair of genetic damage requires the diploid state. As mentioned in Figure 2, when the ploidy level can be varied by experimental conditions, diploid cells are generally much more resistant to DNA damage than haploid cells (for examples in bacteria see Krasin and Hutchinson 1977 and in yeast see Chlebowicz and Jachymchzyk 1979). In the yeast *Saccharomyces cerevisiae*, Herskowitz states that "...diploid cells are better than haploid cells in coping with DNA damage..." (Herskowitz 1988, p. 544, and references cited therein). Similarly, the phenomenon of multiplicity reactivation, whereby phage are much more resistant to DNA damage if they multiply infect a cell than if the singly infect a cell is known to result from the recombinational repair of damages that occurs in the multiply infected (diploid) state (see, for example, the classic paper of Luria 1947).

While haploids do not enjoy the benefit of recombinational repair, they may benefit from being smaller in size than diploids. Research on current unicellular organisms, like yeast, supports the view that diploids are usually larger than haploids. For example, Herskowitz says that diploids "...have a volume nearly twice that of haploids..." (1988, p. 537). Mortimer (1958) found that cell volume scales linearly with ploidy from haploid, diploid up to hexaploid cells. If we assume that diploid cells have twice the volume, they would have about 1.59 times the surface area of a haploid cell. For this reason, we include a parameter p in our models representing the size difference between diploid and haploid cells, and set $p = 1.59$ in many of the studies we report here. Environmental conditions can affect the relationship between ploidy and cell volume. Adams & Hansche (1974) and Weiss et al. (1975) found that the size and metabolism of yeast cells are complex functions of resource limitations. In the extreme case of carbon starvation, the diploids and haploids differed only in the amount of DNA in their cell bodies: the diploids reacted to this environment by becoming smaller. All other measured cell constituents were equal, including quantities of RNA, cell volume, and surface area. If we wanted to represent this situation, we would set $p = 1$ in the models studied below.

Haploids may also require fewer nutrients and resources to replicate than do diploids, since haploids have just one copy of the genome. For simple unicellular organisms, the energy, resources and nutrients involved in making and maintaing the genome may be a sizable fraction of the total resource budget. For this reason, haploids may be more efficient replicators than diploids. For example, in the isomorphic brown alga *Gracilaria verrucosa*, Destombe et al. (1993) found that haploid individuals grew faster than diploids in regular sea-water, while diploid individuals grew faster than haploids in enriched sea-water. Similar results have been reported for haploid and diploid strains of the yeast *Saccharomyces cerevisiae*. (Adams and Hansche 1974).

Asexual Haploid Reproduction

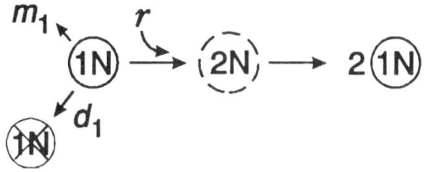

Pure Sex

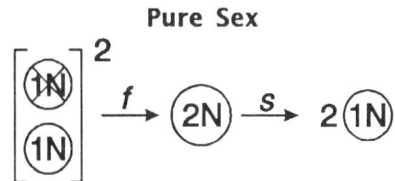

Asexual Diploid Reproduction

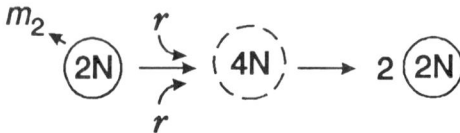

Figure 4. Model Life Histories for the Origin of Diploidy and Sex (Long and Michod 1993, Michod and Long 1993). Haploid cells are sensitive to genetic damage while diploid cells are resistant to damage. The parameters are: m for mortality, s for splitting, f for fusion, and d for damage. The subscripts 1 or 2 refer to the ploidy level, so that, for example, m_1 and m_2 are the rates of cell death to haploid cells and diploid cells, respectively. Diploid cells are assumed to be completely resistant to damage so $d_2 = 0$. The variable N refers to the number of copies of each chromosome, so, for example, $2N$ indicates diploidy. One or two resource "packets", r, are required for reproduction according to whether the cell is haploid, or diploid, respectively. Gene death and mortality occur during sex but are left out of the figure for simplicity. Cell stages not explicitly included in the model are represented by dashed lines. Gene damaged cells are indicated by a crossed-out cell. Sex may or may not involve gene damaged cells. To be concrete a mating between a gene damaged cell and a healthy cell is shown. Repair occurs in the fused diploid state so the two daughter cells are healthy. The $[\ ]^2$ in the Figure indicates the assumption of mass action used in the models, according to which the density of matings between two haploid cell types is equal to the product of the respective haploid cell densities. Two mating systems were considered random mating and damage-induced mating, according to which a damaged cell initiates mating. For reasons of space we only present the results of damage-induced mating here.

3.2. Model

Using the ideas just mentioned, Long and Michod (1994) and Michod and Long (1994) considered the problem of the origin of the sexual life cycle from the component haploid and diploid cell stages. Three basic life cycles are depicted in Figure 4, along with the parameters that underlie our analysis. Two asexual life cycles exist according to whether the cells are diploid or haploid. The sexual life cycle involves fusion of haploids to form diploids and splitting of diploids to form haploids. Sex may or may not involve replication of DNA in the fused state. Thus we distinguish between "Pure Sex" and "Sex Coupled with Reproduction." In pure sex (Figure 4), there is no cell reproduction in the fused diploid state (just splitting). Under pure sex, just haploid cells reproduce. When sex becomes coupled with reproduction, DNA replication occurs in the diploid state. For example in meiosis, there is usually a premeiotic replication in the diploid state yielding four haploid cells after splitting is complete. Sex coupled with reproduction will not be discussed here, although the question of why sex became associated with reproduction is fundamental to understanding the further evolution of sexual life cycles. A preliminary model of this question is given in Long and Michod (1994).

Bell (this volume) points out that no known organisms practice pure sex. The sexual cycle always exists in combination with either haploid or diploid vegetative growth. He argues that sex with haploid vegetative growth was probably the ancestor of all other sexual cycles. For this reason, Long and Michod (1994) and Michod and Long (1994) developed mathematical models of a haploid cell that replicates asexually (as in Figure 4 top panel), while at times undergoing cell fusion and splitting (as in Figure 4, middle panel). Consequently, when we refer to the origin of "sex" we mean the origin of a life cycle involving both the sex and asexual haploid growth. There is an intrinsic cost to sex in such models, since cells undergoing fusion cannot be reproducing. The parameters in Figure 4 are f for fusion, s for splitting, m for cell mortality and d for gene death. There is also a birth parameter, b, that we scale to one as explained below. Further assumptions are given in the caption to Figure 4.

There are two sources of mortality included in the model, gene death and cell death. Gene death may lead to cell death if the function of a needed gene is lost. However, damaged genes may exist in an otherwise functional cell, if the gene is unexpressed or if it has already been expressed. Such damages would interfere with replication (or transcription) should the cell reproduce, but the cell would otherwise appear functional. Gene dead cells may not replicate, but they may undergo other life history events such as fusion if they are sexual. Gene-damaged cells are termed gene dead and are kept track of in the analysis, since they can be repaired and converted into healthy cells. Cells that are cell-dead revert to resources, while gene dead cells remain in the population until they become cell-dead or are repaired. For these reasons, the two sources of mortality, gene death and cell death, are represented separately in the models.

Differences between haploid and diploid cells in their respective rates of gene death and cell mortality are assumed. Damages are assumed to be immediately repaired in diploid cells, while haploid cells are susceptible to DNA damage ($d_1 = d > d_2 = 0$). Cell mortality may also differ between diploid and haploid cells. The results presented below are for the case in which cell mortality is caused by disruption of the cell membrane and so may be a function of cell size. Diploid cells are often larger than haploid cells and so may suffer increased cell mortality.

For the moment, we ignore recessive deleterious mutations, although we believe them to be of fundamental importance to understanding further evolution of the sexual cycle. Since the diploid stage is transient, we believe it unlikely that mutations will accumulate to any appreciable extent. However, Kondrashov (this volume) has pointed out that deleterious mutations may accumulate to significant numbers, even in haploid populations, if extreme forms of truncation selection operate at a threshold number of mutations. Whether such extreme forms of truncation selection exist is at present unknown, but it would be interesting to incorporate both forms of error, mutation and damage, in models similar to those studied here.

An important assumption of the models studied by Long and Michod (1994) and Michod and Long (1994) is that cell replication requires nucleotide resources. We imagine that nucleotide resources are organized into "packets" that contain enough resources to replicate one haploid genome. We assumed that a haploid cell must encounter a single resource packet to reproduce, while a diploid cell must encounter two resource packets. We assumed mass action kinetics. One consequence of mass action kinetics, is that the replication rate of asexual diploids must be lower than the replication rate of asexual haploids, *all other factors being equal*. Thus, there is a baseline disadvantage to diploid cell replication relative to haploid cell replication in our models. We showed (Long and Michod 1993), that the effective diploid birth rate is $\dfrac{1}{\sqrt{2}+1}$ ($\approx 2/3$) the rate of the haploid, in the absence of any size or other intrinsic differences between haploid and diploid cells. This baseline difference in effective birth rates between diploid and haploid cells results directly and necessarily from the our assumption of mass action kinetics. We made additional assumptions that ameliorated this baseline disadvantage of diploidy in replication, so that the ratio of diploid to haploid births was approximately 3/4 in the model whose results are reported here ($p/2 = 1.59/2 \approx 3/4$, as in equation (1) below).

We consider three basic cell types: the asexual haploid, denoted A_H , the sexual haploid, denoted S_H, and the asexual diploid, denoted A_D , The sexual and asexual haploid species may either be healthy or damaged. The sexual haploid may also exist in a fused diploid state. This gives a total of six cell types, three for the sexual haploid, two for the asexual haploid and one for the asexual diploid. The variables representing the densities of these different cell types are summarized in Table 1. Resources that are not contained in cells are free to be used in making new cells. We let r be the density of free resources.

In order to make the equations dimensionless, we divide all equations by the birth parameter, b (we have no intention of setting b to zero), and redefine our other parameters as follows: $t \equiv bt$, and $f \equiv \dfrac{f}{b}, s \equiv \dfrac{s}{b}, m \equiv \dfrac{m}{b}, d \equiv \dfrac{d}{b}$.

Table 1. Cell Density Variables in Model.

	healthy	damaged
A_H; asexual haploid	x_a	z_a
S_H; sexual haploid	x	z
S_D; sexual diploid	y	

With this scaling, all variables are now dimensionless. Parameters, that used to be rates, now represent ratios of rates. Since our equations are autonomous, these changes do not have any significant effect on the form of the equations; but we have achieved a reduction in complexity, from five free parameters to four. We consider here just the case of damage-induced mating, in which only damaged cells initiate mating, with either damaged or undamaged partners. The case of random mating is discussed in Michod and Long (1994). The equations take the final form for damage-induced mating:

$$x' = x(r - m - d) + 2sy - fxz$$
$$y' = fz(z + x) - y(pm + s)$$
$$z' = dx - z[m + f(2z + x)]$$

(1)
$$w' = wp(\frac{r}{2} - m)$$

$$x_a^{'} = x_a(r - m - d)$$

$$z_a^{'} = dx_a - mz_a$$

where free genetic resources are assumed to take the following linear form,

(2) $r = 1 - x - 2y - z - 2w - x_a - z_a.$

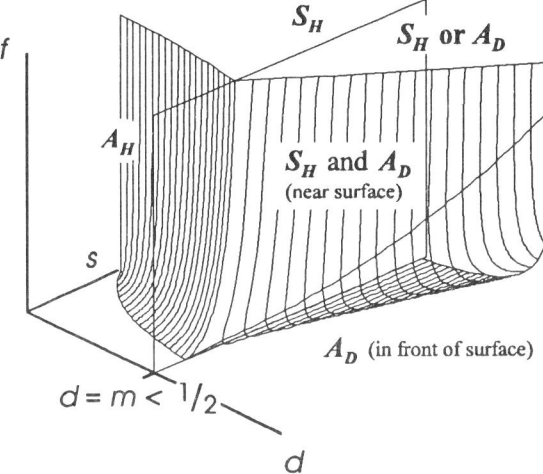

Figure 5. Bifurcation Surfaces for Three-Way Competition (m < 0.5). The winning species is indicated for the different regions: A_H = asexual haploid, S_H = sexual haploid, A_D = asexual diploid. To the left of the $d = m$ plane only haploids win, with the asexual winning in front of the surface and the sexual winning behind the surface (where the splitting rate is high). To the right of the $d = m$ plane the winner is either the diploid or the sexual haploid or both may coexist. Behind the surface (still considering $d > m$) either the sexual haploid or the diploid wins, depending on initial conditions. That is, both the sexual and the diploid are stable when common, but neither can invade when rare. In front of the surface the diploid wins. However, the outcome of competition is more complex near the surface where coexistence of asexual diploids and sexual haploids is possible indicated by "S_H and A_D" in the figure. The surface has an interesting shape as shown in Figure 6 and described in the Appendix of Long and Michod (1994). The intersection of the three "surfaces" is discussed in Long and Michod (1994) and is the only potential site at which all three populations could coexist. Its instability rules this out, however. Figure adopted from Long and Michod (1994).

3.3. Results

We present the results of Long and Michod (1994) in terms of bifurcation diagrams that partition the parameter space into regions occupied by species of the three different types of life cycles: the asexual haploid, A_H ; the sexual haploid, S_H; and the asexual diploid, A_D. In Figure 5, we show the outcome for a particular value of cell mortality, $m = 0.4$, a value of m that lies in the region where the asexual diploid may persist, that is in the region $m < \frac{1}{2}$. Assuming different mortality values in this region, does not change qualitatively the conclusions we draw below.

We can summarize the results depicted in Figures 5 and 6 as follows. Recall that the parameters are scaled to the birth rate, b. If $m > \frac{1}{2}$, the asexual diploid is extinct. If $m + d > 1$, the asexual haploids go extinct. If $d < m < \frac{1}{2}$, then either the sexual wins, or the asexual haploid

wins depending on s. If $d > m < \frac{1}{2}$, then either the sexual or asexual diploid win depending on initial conditions and the parameters of the sexual cycle f and s.

These results suggest plausible scenarios for the origin of sex and diploidy. If we take as primitive the asexual haploid cycle, then the stability results in Figure 5 suggest that sex would evolve before asexual diploidy. This is because the asexual diploid is not competitive in the region of damage for which the asexual haploid can exist ($d < m < 1/2$), however the sexual haploid

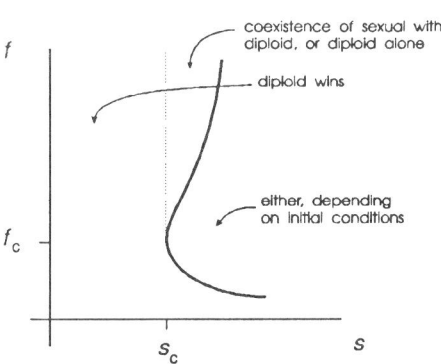

Figure 6. Slice of Bifurcation Surface in Figure 5. See text for explanation and Appendix Figure 2A of Long and Michod (1994) for more technical detail. Figure adopted from Long and Michod (1994).

may win in this region if its splitting rate exceeds the threshold shown by the nearly vertical curtain in Figure 5 of about $s = pm$ (Long and Michod 1993). This becomes more likely as d begins increasing from some initial value, $d_i < m < 1/2$, since the splitting threshold required for sex to be competitive decreases. As d increases above $d = m$, then diploidy becomes more competitive and would resist invasion by sex. However, if the sexual cycle has already been established (when in competition with asexual haploidy alone), sex will continue to be maintained and resist invasion by asexual diploidy (region behind surface in Figure 5 for $d > m$).

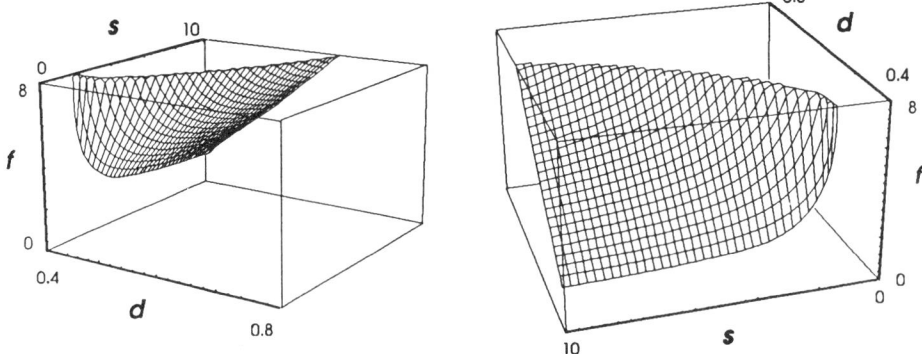

Figure 7. Persistence of Sexual for High Damage and High Mortality ($d + m > 1$, $m = 0.6 > 0.5$). Surface allowing continued existence of sexual is shown for $m = 0.6$. The asexual haploid cannot survive in this region since $d + m > 1$. The asexual diploid cannot survive in this region either, since $m = 0.6 > 0.5$ (Michod and Long 1993). Only the sexual can persist in this range ($d + m > 1$, $m > 0.5$), and only if the sexual parameters (s and f) are sufficiently strong. Points above the surface represent parameter values for which the sexual can still maintain a positive rate of increase (given sufficiently sized initial populations). The two graphs contain identical information but are shown from different perspectives. Figure adopted from Michod and Long (1994).

For high mortality and high damage ($m > \frac{1}{2}$ and $d + m > 1$) the sexual cycle may still persist even though its component cell types would go extinct if they were to reproduce asexually. The region of persistence of the sexual for high damage and high mortality is shown in Figure 7 for $m = 0.6$. The asexual haploid cannot survive in the region shown, since $d + m > 1$. The asexual diploid cannot survive in this region either, since $m = 0.6 > 0.5$ (Long and Michod 1993). Only the sexual can persist in this range, and only if the sexual parameters (s and f) lie in the specified region. Points above the surface represent parameter values for which the sexual can still maintain a positive rate of increase (assuming sufficient initial densities).

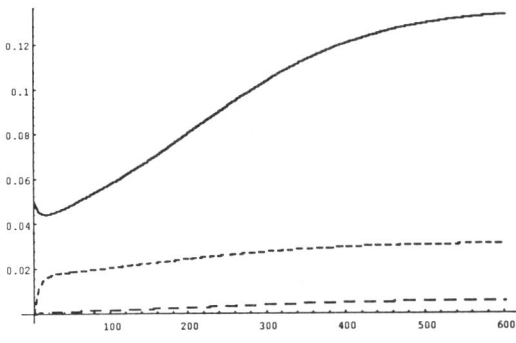

Figure 8. Increase of Sex in Extreme Environments. The abscissa is generation and the ordinate is frequency. The solid (upper) curve is the density of healthy cells, the dotted (middle) curve is gene-dead cells and the dashed curve is fused diploid cells. See text for explanation.

Although the sexual cycle can maintain itself in these extreme environments it certainly does not flourish there--reaching densities of between 5-20% depending on the severity of the environment as measured by d and m. That is, most resources (80-95%) are free and not in cells. For example, in Figure 8 is shown the increase of the sexual cycle for the parameter values $m = 0.6, d = 0.5, f = 8, s = 6, p = 1.54$ beginning from initial population densities of $\left(x_0 = 0.05, z_0 = 0, y_0 = 0\right)$. There is an initial decrease in healthy cells as damaged and fused cells establish themselves, but soon healthy cells begin increasing up to the locally stable equilibrium of $\left(\hat{x} = 0.1359, \hat{z} = 0.0311, \hat{y} = 0.0060\right)$. For lower initial densities the sexual cycle has difficulty establishing itself and for an initial density of about $\left(x_0 = 0.01, z_0 = 0, y_0 = 0\right)$ cannot increase at all. The asexual haploid and diploid cycles go extinct no matter what their initial densities.

We now turn to a basic problem that sexual cells have especially when rare: encountering a mate. If sexual fusions require random encounters between sexual cells, such encounters are extremely rare when sex first originates. This is because the likelihood of encounter and fusion is assumed to be proportional to the product of the population densities of the mating cells. These same non-linearities that make the expansion of sex from extreme rarity difficult, also tend to make sex stable when it is common (Bernstein et al 1985, Michod 1991).

However, this cost of rarity for sex assumes the sexual and asexual populations to be reproductively isolated from one another. In Michod and Long (1994) sexual cells were allowed to mate with either asexual and other sexual cells, as if in a Mendelian-like population. It may seem unreasonable to assume that a sexual cell could "coerce" an asexual cell into mating. Yet, a similar assumption is made in models of infectious transfer of selfish elements (see, for example, Hickey and Rose 1988). Indeed, such coercion is precisely what the F-plasmid orchestrates during conjugation in the bacterium $E.\ coli$. In infectious transfer models, it is commonly assumed that the infectious element causes matings between cells, one cell that has the element and one cell that doesn't have the infectious element. In Michod and Long (1994), sexual and asexual cells are similar, respectively, to cells with or without the infectious element. When matings occur between a sexual and an asexual cell, we assume that a healthy sexual and an healthy asexual cell type is produced upon splitting.

Allowing sexual cells to mate with asexual cells dramatically increases the selective advantage of sex when rare, even though damaged asexual cells may also be repaired. As shown in Michod and Long (1994), asexuality is often unstable to sex when mixed sexual-asexual matings occur. In the case of the Mendelian-like population genetics model studied, the advantages of sex are of first order. Even when sexual cells are rare, there are asexual cells to mate with. In the inter-species competition model studied by Long and Michod (1994) and reported above, the advantages of sex are second-order when sexual cells are rare, since sexual cells can only mate with other sexual cells. Furthermore, in the inter-species model, many of those infrequent matings that occur when population densities are rare tie up healthy sexual cells and prevent them from reproducing until splitting occurs. However, in the Mendelian-like model when sex is rare, most matings are with asexual cells. Consequently, healthy sexual cells tend not to be tied up repairing damaged cells. For this reason, the splitting rate is not as critical an issue in the Mendelian-like model as it was in the group selection model studied by Long and Michod (1994).

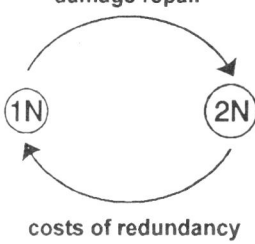

damage repair

costs of redundancy

Figure 9. Forces of Selection on the Origin of the Sexual Cycle. See text for explanation.

In summary, in Figure 9 is represented the basic forces of selection hypothesized here for the origin of the sexual cycle. The diploid stage is advantageous because of repair of DNA damage, while the haploid stage is advantageous because it avoids the costs of replication in the diploid stage. These costs of redundancy stem primarily from the cost of nucleotide resources. While the costs of accruing nucleotides may be relevant to the origin of the sexual cycle, they likely became of decreasing importance as the diploid stage of the cycle became dominant--especially when multicellular organisms evolved. The costs of DNA are probably a small part of the resource budget in multicellular organisms. At this point, we now argue that individual heterozygosity may have helped maintain the outcrossing aspects of the sexual cycle.

4. Heterozygosity and the Further Evolution of Outcrossing

4.1. Transition to diploid sexual cycle

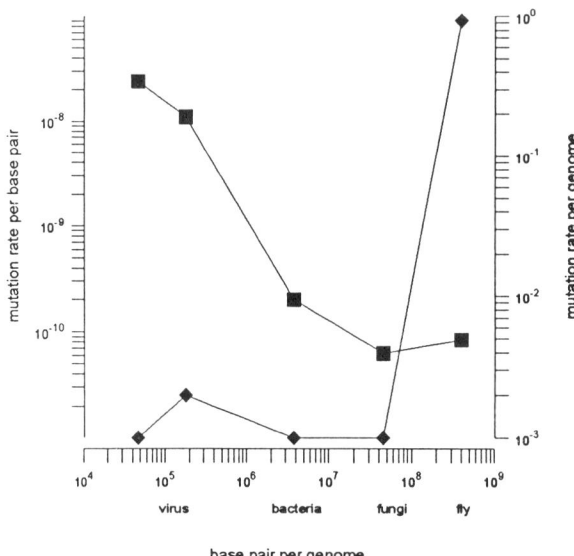

Figure 10. **Mutation Load as a Function of Genome Size.** Data taken from Drake 1974. ■ = mutation rate per base pair replication; ◆ = mutation rate per genome per generation.

The haploid sexual cycle is characterized by asexual replication of haploid cells and temporary fusion of haploids to create diploids that quickly split into haploids. The diploid sexual cycle is the reverse: most asexual replication occurs in the diploid stage and the haploid stage is transient. As hypothesized in the above models, diploidy was initially a transient stage that functioned in the recombinational repair of DNA damage. It was transient because of the efficiency of haploid replication (Figure 9). At some point in evolution, the diploid stage became dominant. Why? Increasingly complex life forms required more and more genetic information. As the genome size expanded during evolution from about 10,000 base pairs (as in simple viruses) to about 10 million base pairs (as in fungi such as *Neurospora*), the rate of mutation per replication of the genome was kept approximately constant, at about 0.001 (Figure 10, diamonds). This was accomplished, by increasing the fidelity of DNA replication, in other words, by decreasing the mutation rate per base pair to about 1 mistake in 10 billion base pairs replicated (Figure 10, squares). It seems that further increases in the fidelity of DNA replication were too costly.

For Drosophila the mutation rate per genome jumps 3 orders of magnitude to about 1 new mutation per generation. This increase is due to the increase in genome size and the increase in the number of cell divisions per generation. Drosophila is the only sexual diploid in the data. It appears that as the genome size increased from about 10 million base pairs (fungi) to about a billion base pairs (Drosophila), it was necessary for the diploid stage to become dominant, probably as a result of the masking of recessive, or nearly recessive, mutations that occurs during the diploid stage.

Once the diploid stage of the sexual cycle became dominant and the haploid stage transient, fusion (outcrossing) was no longer needed for repair of damages--since cells were diploid most of the time anyway. In principle, recombinational repair should occur in the diploid (asexually reproducing) state just as it occurred in the fused diploid state of the sexual cycle. The costs of genetic redundancy, that were a selective factor in the origin of the sexual cycle, were unlikely to be of continued importance as the asexual diploid stage of the cycle became enhanced.

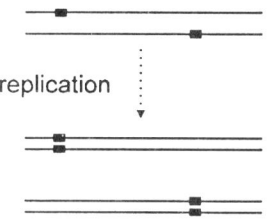

Figure 10. Replication of Heterozygous Mutations. Two heterozygous mutations in a diploid cell (top) are replicated to form two pairs of sister chromosomes. Figure applies to any heterozygous locus.

The question then becomes, why is outcrossing maintained in diploids? In this section we give a qualitative overview of the role of heterozygosity and homozygosity in the maintenance of outcrossing. In the next section, we develop heuristic principles that underlie quantitative models of the issues discussed. We continue to hypothesize that recombination occurs during the diploid state for the function of repair of genetic damage. The remaining issue is simply whether the two DNA molecules in a diploid cell come from the same individual, as in recombination during selfing, or from another individual in a previous generation, as in outcrossing. Up to this point we have been considering only damages. But the other kind of genetic error, mutation, now becomes of central importance. As the diploid stage of the sexual cycle became dominant, deleterious recessive, or nearly recessive, mutations accumulated in the genome masked in heterozygous state. They now threaten survival if they are unmasked by becoming homozygous. In addition, overdominant loci that are expressed during the diploid stage lower fitness when made homozygous.

In Figure 10, we show the two chromosomes in a diploid cell after they have replicated to form two pairs of sister chromosomes. Mutations heterozygous before replication are now present in each chromosome of each sister-chromosome pair. By picking one chromosome from each sister-chromosome pair (Figure 11 left; as is done during mitosis) these mutations are kept from being unmasked and are maintained heterozygous. However, the situation is more involved if recombination has repaired a damage as shown in Figure 11 (right).

In Figure 11 (right), an unrepaired double-stranded damage is assumed to be present in one of the chromosomes. After replication, a gap is formed in the sister chromosome at the site of the damage. Recombination with one of the chromosomes from the other sister-chromosome pair repairs the damage. However, loci flanking the site of recombination may now exchange position if crossing-over is associated with the repair event (recall Figure 3). Since the occurrence of crossing-over is a random event depending on the isomerization of the Holliday junctions in structural intermediates in the recombination pathway, there is no way of being sure where the mutations are after recombinational repair. Consequently there is a high likelihood of making deleterious mutations homozygous, if chromosomes are taken from the same individual (as, for example, in selfing) to produce a daughter cell. However, if a chromosome is taken from an unrelated mate, as in outcrossing, heterozygosity and masking of mutations will be maintained.

There are many reproductive systems in which outcrossing has been abandoned. In other words, evolution has repeatedly taken species down the path indicated by the "without" arrow in Figure 11 (right diagram). Selfing in plants and automixis in animals are such examples in which the outcrossing aspects of sex have been lost. Another possible example would be some kind of asexual diploidy with recombinational repair but no haploid stage. All such systems are "closed" in the sense that recombination occurs between chromosomes taken from the same individual in a previous generation. It is clear that the barrier to loss of outcrossing is not absolute. We will take up this issue in the next section.

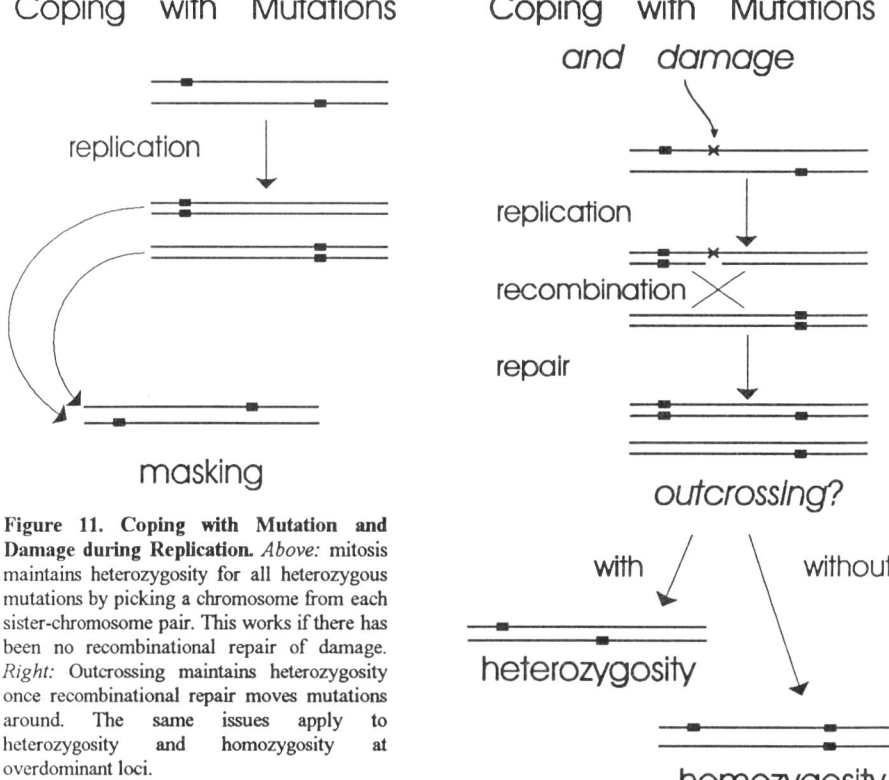

Figure 11. **Coping with Mutation and Damage during Replication.** *Above:* mitosis maintains heterozygosity for all heterozygous mutations by picking a chromosome from each sister-chromosome pair. This works if there has been no recombinational repair of damage. *Right:* Outcrossing maintains heterozygosity once recombinational repair moves mutations around. The same issues apply to heterozygosity and homozygosity at overdominant loci.

One may ask why evolution did not re-invent a generalized recombination repair mechanism with no crossing-over (that is, somehow bias the outcome of Figure 3 to the no crossing-over products). Specialized recombination systems with no crossing-over are known to occur in, for example, mating type switching in yeast. Such a general recombination mechanism would avoid the need of outcrossing to maintain heterozygosity (so long as recombinational repair did not occur at the site of mutations). The answer to this question may simply be history and the fact that two complicated changes are needed simultaneously for either to be beneficial (Bernstein, Hopf and Michod 1988). As we have seen, there is accumulating evidence that the recombination mechanism is ancient and highly conserved throughout living systems. In the models discussed previously, outcrossing originated to provide for diploidy while avoiding the costs of replication in the diploid state. Once originated, outcrossing would provide for heterozygosity and masking of recessive mutations that accumulated as the diploid state became more pronounced. There is no advantage to re-inventing a molecular recombination system, even if it were possible to do so, unless outcrossing is simultaneously abandoned (outcrossing is already effectively maintaining heterozygosity). There is a great penalty to pay if outcrossing is abandoned without re-inventing the recombination system to have no crossing-over, because of the homozygosity that would occur. Thus, both a new recombination system and abandonment of outcrossing must occur simultaneously for either to be advantageous. And, of course, this would be very unlikely. It seems that history has left us with a less than optimal, but workable, solution: the diploid sexual cycle.

4.2. Understanding the evolution of selfing and outcrossing

In the previous section, we asked why diploid species with recombination do not abandon outcrossing. We presented in broad qualitative terms the hypothesis that outcrossing is maintained to keep loci heterozygous, thereby avoiding the detrimental effects of either expressing recessive deleterious mutations or homozygosity at overdominant loci. In recent years, there has been considerable theoretical work on the role of heterozygosity in maintaining the outcrossing aspects of sex, especially in competition with regards extreme forms of inbreeding such as sib-mating and selfing. The question of the evolution of outcrossing and inbreeding systems of mating has been of special interest to students of plant mating systems both from the theoretical and empirical points of view. Extensive theoretical work has been done employing computer simulation of multilocus models and mathematical analysis of one- or two-locus models. Most of this work studies the evolution of outcrossing (or outbreeding) and explicit forms of inbreeding, such as selfing or sib-mating. In this Section we focus on the role of individual heterozygosity in the evolution of selfing and outcrossing (random mating) as a kind of test case of the general hypothesis presented in Section 4.1--keeping in mind the hypothesis discussed in Section 2 that the recombinational aspects of sex function in DNA repair. When viewed from this perspective, the problem of the evolution of outcrossing and selfing becomes of central relevance to the problem of the evolution of sex.

4.2.1. Inbreeding depression and the "ripple" effect

The concept of inbreeding depression has been a central issue in the study of mating system evolution. Inbreeding depression measures the extent to which the fitness of selfed offspring is lower than the fitness of outcrossed offspring. The recent reevaluation of the role of inbreeding depression in directing evolution of selfing rates has resulted in a number of quantitative studies on the subject. Lande and Schemske (1985), Campbell (1986), Holsinger (1988), Michod and Gayley (1992) and Uyenoyama and Waller (1991a, 1991b, 1991c) have challenged the usefulness of inbreeding depression in understanding mating system evolution, whereas others (Charlesworth and Charlesworth (1990), Charlesworth, Morgan, and Charlesworth (1990)) have sought to rescue the concept.

These studies have used different methods and offered different explanations for their results. Some studies have relied on computer simulations (Holsinger, Charlesworth and Charlesworth (1990), Charlesworth, Morgan, and Charlesworth (1990)), resulting in tables of special-case results. Uyenoyama and Waller (1991a, 1991b, 1991c) have used rigorous mathematical analysis, and thus require a number of simplifying assumptions and obtain results only for initial-increase conditions. Our goal in this section is to step back from the complexity and details of these analyses and show that by considering a few simple concepts we can explain essentially all the qualitative results from the studies in an intuitive framework. In particular, the role of inbreeding depression in evolutionary explanations can be clearly understood.

We elaborate on the fundamental idea, understood by all these authors, that genetic associations develop between mating-system modifier alleles and fitness-determining alleles. The concepts we develop here 1) apply to both heterozygote superiority and mutation-selection balance models; 2) explain, at least qualitatively, the effect of the strength of the modifier allele on conditions for invasion; 3) explain why and when inbreeding depression will correctly predict evolution at a modifier locus; 4) provide more insight than just referring to "genetic associations between the modifier and fitness loci."

We consider the following situation. Individuals in a population produce both male and female function (for example, pollen and ovules). For any fraction of their ovules, they can either open up them up to outside pollen (i.e., outcross), or they can contribute the entire diploid complement (self). We assume that all individuals produce the same pollen output, regardless of their selfing rate (this is commonly referred to as "no pollen discounting"). This final assumption imposes the so-called twofold cost of sex, because individuals lose no pollen success when they self, yet gain a twofold transmission advantage in their ovules. In practice, pollen discounting may reduce the cost of outcrossing so that the evolution of outcrossing is made easier. However, we wish to study outcrossing in a worst case scenario when the costs of outcrossing are large.

In the face of this large cost of outcrossing, it is hypothesized that the benefit of outcrossing comes from the production of more-fit offspring through increased heterozygosity. Heterozygosity is assumed to be advantageous because of heterozygote superiority, or masking of deleterious recessive mutations. A basic measure of the difference in fitness of selfed and outcrossed offspring is inbreeding depression, defined as one minus the ratio of the fitness of selfed to outcrossed offspring. By this measure, it would seem that the break-even point for outcrossing and selfing is that outcrossed offspring have twice the fitness of selfed ones, an inbreeding depression of 1/2. We refer to this as the "inbreeding depression criterion" for predicting the success of mating system modifiers. The familiar value of 1/2 is specific to the case where there is no pollen discounting.

Inbreeding depression can be determined empirically by measuring the fitnesses of a sample of offspring known to have been produced by selfing and outcrossing. In genetic models, it is usually calculated from the genotype frequencies and fitness parameters, but it is easier to think of it as the result of a thought experiment: take all the genotypes in the population, self them, and measure the fitness of the offspring, then take all the genotypes in the population, outcross them, and measure the fitness of those offspring. An important point here is that inbreeding depression is a population-average quantity.

Now introduce a modifier allele that alters an individual's selfing rate. Say, for this example, that the initial population is pure outcrossing and that the modifier causes its carriers to self entirely. Furthermore, assume that the population is undergoing recurrent mutation to deleterious recessive fitness alleles and thus individuals carry a complement of these alleles in the heterozygous (masked) state. We could calculate an inbreeding depression for this population by mathematically performing our thought experiment. Depending on the mutation rate and other parameters, it could easily be much greater than 1/2, due to unmasking of the recessive deleterious alleles in selfed offspring.

The modifier takes a big hit in offspring fitness in the first generation due to unmasking. This fitness cost is by definition what inbreeding depression measures. But as a direct result of this big hit, the modifier has purged some of the deleterious mutations from its background. In the next generation, the surviving carriers of the modifier will self, and the fitness consequences of selfing for them will be very different from the previous generation, due to the change in their genetic background. Yet the inbreeding depression for the population is unchanged (the modifier's frequency is still negligible), and thus it no longer has any bearing on the future course of evolution of the modifier.

We refer to the effect a mating-system modifier allele has on its own genetic background as its "ripple effect". Inbreeding depression measures the fitness consequences only for the first generation of a modifier's existence, when its genetic background is the same as the rest of the population. Thus, in cases where a modifier will create a small ripple, we expect to find that the inbreeding depression criterion is close to being a correct predictor of the invasion success of modifiers. Conversely, in cases where the modifier creates a large ripple around itself, we expect that inbreeding depression will not be a correct predictor.

Note that this insight about the ripple effect is not the same as the oft-repeated observation that inbreeding depression changes as the mating system evolves. Certainly, this observation is correct and is part of the reason that many simple predictions based on inbreeding depression fail, but whether inbreeding depression is allowed to vary with evolution or not, it is still a population-average measure. It is essential to consider the environment that the modifier creates for itself--in effect, it has its own inbreeding depression.

We now show how an analysis based on the ripple effect can predict essentially all the qualitative results about modifier invasion conditions known form the studies cited so far. We discuss separately the cases where fitness variation is maintained by heterozygote superiority and by recurrent deleterious recessive mutation.

4.2.2. Heterozygote superiority

We restrict attention to the case where there is a single overdominant fitness locus, with equal homozygote fitnesses. In this case, inbreeding depression is always less than 1/2 for any initial selfing rate and fitness values (Ziehe and Roberds (1989), Uyenoyama and Waller (1991b), and a polymorphism is maintained at all selfing rates. Thus, according to the inbreeding depression criterion, we would expect increased-selfing modifiers to always invade, and increased-outcrossing modifiers to always fail. This is not the case.

Consider first the case of modifiers with a higher outcrossing rate than the current value in the population (increased-outcrossing modifiers). For the purposes of our reasoning, it is convenient to think of an individual's outcrossing probability as being realized as a cycle of generations of pure selfing and pure outcrossing. For example, an outcrossing rate of 1/3 corresponds to a cycle of two generations of selfing followed by one of outcrossing. Furthermore, we imagine that the initial population (all those who do not carry the modifier) is going through this cycle in synchronization. An individual with a higher rate of, say, outcrossing, will outcross in all the generations that the population outcrosses, and some additional ones as well. Nothing in our arguments depends in any way on such an artificial scenario, but it helps to clarify the consequences of "excess" acts of selfing or outcrossing.

We already know that an act of outcrossing never pays for itself in the generation it occurs, because the increased heterozygosity of outcrossed offspring never makes up for the transmission cost of not selfing (this from the inbreeding depression value). However, by outcrossing you have injected increased heterozygosity into your lineage. If you go back to selfing right away, you will have many generations of "payback" where you get to enjoy the benefits of increased heterozygosity without paying any cost of sex. Your increased heterozygosity will gradually diminish to the level of the rest of the population, but if you have enough generations of selfing before you outcross again, it is possible to more than make up the initial deficit incurred. Therefore, we expect that small-effect increased-outcrossing modifiers may be able to invade. This phenomenon can be seen in the results of Holsinger (1988, Table 2), Charlesworth and Charlesworth (1990, Table 4), Uyenoyama and Waller (1991b), and Michod and Gayley (1992). The size of the modifier is crucial to its invasion chances--outcrossing is never beneficial in itself, but it can be thought of as a strategy for selfers to inject some heterozygosity into their lineage.

Clearly, then, it is advantageous to not outcross again for as long as possible. The outcrossing rate of carriers of the modifier can be broken down into two components: the initial population value and the change due to the modifier. The smaller either of these values is, the easier it will be for a modifier to invade. The advantage of small-effect modifiers can be seen in Table 4 of Charlesworth and Charlesworth (1990) and Michod and Gayley (1992). The advantage of a low initial outcrossing rate is consistent with Uyenoyama and Waller (1991b) Figures 2 and 4. A lower initial outcrossing rate also helps outcrossing modifiers by raising the inbreeding depression (see, for example, Uyenoyama and Waller (1991b)), which means that the initial hit in the first generation of outcrossing is reduced.

In summary, an increased-outcrossing modifier creates a significant ripple around itself, but to take advantage of that ripple, it must refrain from outcrossing again for as long as possible. The ripple effect works to the advantage of outcrossing modifiers, so the inbreeding depression criterion is expected to be too stringent in predicting their invasion conditions.

Now consider the case of increased-selfing modifiers. We know from the inbreeding depression result that, starting from an average genetic background, an act of selfing always has positive fitness consequences. Thus, a selfing modifier "wins" in the first generation. However, as a result of this "excess" act of selfing, the modifier's background is slightly more homozygous than average. Each successive generation of "excess" selfing makes one's background more homozygous than the rest of the population, which is to say that the ripple is detrimental, and it is larger for modifiers of larger effect.

If the initial population is pure outcrossing, then since the mean fitness of a pure selfing population is never less than 1/2 that of pure outcrossing, the ripple never gets bad enough to prevent even complete-selfing modifiers from invading (Uyenoyama and Waller (1991b) equation 14, Michod and Gayley (1992)). However, consider what happens if the initial selfing rate is high. The population is selfing in almost every generation, but by assumption there will be some generation in which the population outcrosses but a carrier of the modifier selfs. This generation of "excess" selfing by the modifier will likely be followed by many more generations of selfing (along with the rest of the population). In every one of those generations, it will have a more homozygous background than the rest of the population. Thus, as the initial selfing rate increases, there may come a point where the advantage a modifier gains with its excess generation of selfing (when everyone else is outcrossing) is outweighed by the accumulated cost of a long series of generations where it has a more homozygous background (demonstrated in Charlesworth and Charlesworth (1990) Table 4, and Uyenoyama and Waller (1991b) Figures 4 and 6).

Therefore, the inbreeding depression criterion is too lenient in predicting increase of selfing modifiers. Just because you win in the first generation doesn't mean you win in the long run, and small-effect modifiers, which take the initial win and then quickly return to the average genetic background, are more likely to invade.

In summary, the ripple effect makes the actual conditions for modifier increase more lenient than the condition derived from considering inbreeding depression for modifiers that increase the outcrossing rate, and more stringent for modifiers that increase the selfing rate. Furthermore, the size of the effect of the modifier is shown to be important, with small-effect modifiers of either type more likely to invade.

4.2.3. Mutation-selection balance

We now assume that fitness variation is maintained by recurrent mutation to deleterious recessive or nearly-recessive alleles. In such a population, the resident inbreeding depression is a decreasing function of the selfing rate; the more selfing, the more purged of mutations the population becomes.

Consider first the case of increased-selfing modifiers. We already described the effect that a large-effect selfing modifier has on its background, and how this can lead to its invasion despite a high inbreeding depression. What about modifiers of small effect? The advantage of purging your genetic background only accrues to those lineages that stick with selfing long enough to where they become better off than the rest of the population. As soon as you outcross, you lose your ripple and effectively have to start over again. In other words, small-effect selfing modifiers create only a small ripple, so we expect the inbreeding depression criterion to be nearly correct for them. The reasoning about a large-effect selfing modifier was taken to its extreme by Lande and Schemske(1985), who point out that a pure-selfing modifier can always invade provided that some of its carriers can survive through the first generations of unmasking and purging of recessives. The ability of large-effect, but not small-effect, selfing modifiers has been demonstrated by Holsinger (1988, Tables 4 and 5), and Charlesworth and Charlesworth (1990, Figure 6).

Now consider increased-outcrossing modifiers. If the mutation rate is not high, then single "excess" acts of outcrossing have little effect on your genetic background (provided, of course, that you are not more purged than the rest of the population). Thus, an increased-outcrossing modifier creates little ripple. Gradually, of course, an outcrossing lineage's background will begin to accumulate recessives to a greater degree than the population, but this occurs on a longer time scale, depending on the mutation rate. If outcrossing is good for you in the first generation, it will be good in subsequent generations, and likewise if it is not good for you. Thus, we expect the inbreeding depression criterion to be nearly correct, and that this should be true regardless of the size of the modifier. This prediction is consistent with the simulation results of Charlesworth, Morgan, and Charlesworth (1990), but needs more study to be conclusively demonstrated.

4.2.4. Strategic Reasoning

The premise of using inbreeding depression to predict the evolution of selfing rates assumes the "survival of the fittest" paradigm for natural selection (Michod 1991), in which organism fitness is the sole determinant of evolutionary change. The same is true of all "strategic" approaches to evolution, for example, ESS arguments based on phenotypic considerations. The basic idea is "we can expect phenotype 1 to increase in competition with phenotype 2 when the fitness of phenotype 1 is greater than the fitness of phenotype 2". As we have sought to demonstrate in the past (Gayley and Michod 1990, Gayley 1993), this type of reasoning is only valid when a consistent fitness effect accrues for all performances of the two behaviors. More general problems with the "survival of the fittest" approach are discussed by Michod (1991) and Byerly and Michod (1991).

In the case of the evolution of selfing and outcrossing, we have seen that due to the effect that mating-system modifiers have on their own genetic background, different mating-system genotypes experience different fitness consequences for their acts. Thus there is no meaningful "average fitness effect of selfing" that can be applied to every act of selfing, and we expect reasoning based entirely on average fitnesses of phenotypes to be incorrect.

This insight makes us more comfortable with two possibilities that have been demonstrated for the heterozygote-superiority case by computer simulations (Holsinger 1988, Charlesworth and Charlesworth (1990)) and analysis of stability conditions (Uyenoyama and Waller (1991b)): the existence of selfing rates that resist invasion by all small-effect modifiers, whether they increase or decrease the selfing rate, and also those that allow invasion of modifiers of either type. These states defy naive intuition because we know that either outcrossing or selfing is "better", so why don't modifiers that increase the better strategy increase? The answer is, of course, that the increase of a selfing modifier and an outcrossing modifier are not logically complementary processes. Rather, they depend on the specific genetic dynamics induced by the modifier allele, and there is no a priori reason, such as based on phenotype fitness, to expect that only one or the other type should ever be able to invade

5. Conclusions

The sexual cycle involves the alternation of haploid and diploid cell states so that recombination and outcrossing occurs. According to the ideas presented here, molecular recombination during diploidy originated and is maintained for the function of DNA repair. Outcrossing (fusion of haploid cells) originated for the advantage of damage repair in the diploid state, while splitting was favored to reduce the costs of replicating in the diploid state. As the diploid stage of the sexual cycle became dominant these costs of redundancy became less significant. Heterozygosity, either at overdominant loci or at loci harboring recessive mutations, is also advantageous in diploid cells and may serve to maintain the outcrossing aspects of sex.

References

1. Adams, J., & P.E. Hansche. Population Studies in Microorganisms I. Evolution of Diploidy in Saccharomyces cerevisiae. 1974. Genetics 76: 327-338.

2. Bernstein, H., Recombinational repair may be an important function of sexual reproduction, BioScience 33 (1983), 326-331.

3. Bernstein, H., F.A. Hopf, and R.E. Michod, The molecular basis of the evolution of sex. Advances in Genetics, 24 (1987), 323-370.

4. Bernstein, H., F. Hopf, and R.E. Michod, Is meiotic recombination an adaptation for repairing DNA, producing genetic variation, or both? In Michod, R. E. and B. R. Levin (editors) *Evolution of Sex: An Examination of Current Ideas* Sinauer Associates. (1988).

5. Byerly, H. C. and R. E. Michod, Fitness and evolutionary explanation. Biology and Philosophy 6, (1991), 1-22.

6. Campbell, R. B., The interdependence of mating structure and inbreeding depression, Theor. Pop. Biol., 30 (1986), 232-244.

7. Charlesworth, D., M. Morgan, and B. Charlesworth, Inbreeding depression, genetic load, and the evolution of outcrossing rates in a multilocus system with no linkage, Evolution 44, (1990), 1469-1489.

8. Charlseworth, D., and B. Charlesworth, Inbreeding depression with heterozygote advantage and its effect on selection for modifiers changing the outcrossing rate, Evolution 44, (1990), 870-888.

9. Chlebowicz, E. and W. J. Jachymczyk, Repair of MMS-induced DNA double strand breaks in haploid cells of *Saccharomyces cerevisiae*, which requires the presence of a duplicate genome, Molec. Gen. Genet. 167 (1979), 279-286.

10. Cox, M. M., The recA protein as a recombinational repair system, Molecular Microbiology, 5 (1991) 1295-1299.

11. Destombe, C. H. Godin, M. Nocher, S. Richerd and M. Valero, Differences in response between haploid and diploid isomorphic phases of *Gracilaria verrucosa* (Rhodophyta: Gigartinales) exposed to artificial environmental conditions. Hydrobioilogia 260/261 (1993), 131-137.

12. Drake, J.W., The role of mutation in bacterial evolution, Symp. Soc. Gen. Microbiology, 24 (1974) 41-58.

13. Gayley, T., Genetics of kin selection: The role of behavioral inclusive fitness. Am. Nat. 141 (1993), 928-953

14. Gayley, T, and R. Michod, Modification of genetic constraints on frequency-dependent selection, Am. Nat. 136, (1990), 406-427.

15. Herskowitz, I., Life cycle of the budding yeast *Saccharomyces cerevisiae,* Microbiological Reviews 52 (1988), 536-553.

16. Holsinger, K., 1988, Inbreeding depression doesn't matter: the genetic basis of mating system evolution, Evolution 42, 1235-1244.

17. Krasin, F. and F. Hutchinson, Repair of DNA double-strand breaks in *Escherichia coli,* which requires recA function and the presence of a duplicate genome, J. Mol. Biol. 116 (1977), 81-98.

18. Lande, R., and D. Schemske, The evolution of self-fertilization and inbreeding depression in plants. I, Evolution 39, (1985), 24-40.

19. Long, A. and R. E. Michod, Origin of sex for error repair. I. Sex versus Diploidy versus Haploidy? Theor. Popul. Biol. (In Press.)

20. Luria,   S. E., Reactivation of irradiated bacteriophage by transfer of self-reproducing units, Proc. Natl. Acad. Sci. USA 33 (1947), 253-264.

21. Michod, R. E., Sex and evolution, Pages 285-320 in *1990 Lectures in Complex Systems,* Santa Fe Institute Studies in the Sciences of Complexity, Lect. Vol III, Eds Lynn Nadel and Daniel Stein, Addison-Wesley, Redwood City, CA., (1991)

22. Michod, R. E., and B. R. Levin, *The Evolution of Sex: An Examination of Current Ideas,* Sinauer Associates, Sunderland, MA., (1988)

23. Michod, R., and T. Gayley, Masking of mutations and the evolution of sex, Am. Nat. 139, (1992), 706-734.

24. Michod, R. E. and A. Long, Origin of sex for error repair. II. Can sex cope with rarity or extreme environments? Theor. Popul. Biol. (In Press.)

25. Miller, R. V. and T. A. Kokjohn, General microbiology of RecA: environmental and evolutionary significance, Annu. Rev. Microbiol. 44 (1990), 365-394.

26. Ogawa, T., X. Yu, A. Shinohara, and E. H. Egelman, Similarity of the yeast RAD51 filament to the bacterial RecA filament, Science 259 (1993), 1896-1899.

27. Roca, A. I, and M. M. Cox, Crit. Rev. Biochem. Mol. Biol., 25 (1990), 415.

28. Story, R. M., D. K. Bishop, N. Kleckner, and T. A. Steitz, Structural relationship of bacterial RecA proteins to recombination proteins from bacteriophage T4 and yeast, Science 259 (1993), 1892-1896.

29. Szostak, J.W., T.L. Orr-Weaver, R.J. Rothstein and F.W. Stahl, The double-strand-break repair model for recombination. Cell. 33 (1983), 25-35.

30. Uyenoyama, M., and D. Waller, 1991a, Coevolution of self-fertilization and inbreeding depression. I. Mutation-selection balance at one and two loci, Theor. Pop. Biol. 40, 14-46.

31. Uyenoyama, M., and D. Waller, 1991b, Coevolution of self-fertilization and inbreeding depression. I. Symmetric overdominance in viability, Theor. Pop. Biol. 40, 47-77.

32. Uyenoyama, M., and D. Waller, 1991c, Coevolution of self-fertilization and inbreeding depression. I. Homozygous lethal mutations at multiple loci, Theor. Pop. Biol. 40, 173-210.

33. Valero, M., S. Richerd, V. Perrot and C. Destombe, Evolution of alternation of haploid and diploid phases in life cycles. Trends Ecology and Evolution. 7 (1991), 25-29

34. Wolfram Research, Inc., 1992, Mathematica, Wolfram Research, Champaign.

35. Ziehe, M., and J. Roberds, 1989, Inbreeding depression due to overdominance in partially self-fertilizing plant populations, Genetics 121, 861-868.

Ecology and Evolutionary Biology, University of Arizona, Tucson, AZ 85721.

Current addresses: Michod is at the Dept. of Ecology and Evolutionary Biology, University of Arizona, Tucson, AZ 85721. Gayley is at Wolfram Research, Champaign, IL.

E-mail addresses: michod@ccit.arizona.edu and tgayley@wri.com

Lectures on Mathematics in the Life Sciences
Volume **25**, 1994

Experimental Approaches to the Evolution of Life Cycles

VÉRONIQUE PERROT

ABSTRACT. The life cycle is the central character of an organism. It shows wide variation, but its evolution is still poorly understood. Different cycles differ with respect to the time an organism spends as a haploid or as a diploid. It is of primary interest to understand the differences between haploids and diploids. In this paper I review the hypotheses which have been proposed to explain these differences, and the experiments which have been performed to test some of them. From the available data, it seems that (1) diploidy is an efficient protection in a highly mutagenic environment, and that (2) diploids seem to be favored in nutrient-rich environments and haploids in nutrient-poor environments. The experimental evidence supporting the latter conclusion is still scarce, and there is need for more data to confirm it.

Until now, we have not managed to compare different populations of the same organism following different life cycles. We need to make this comparison in order to understand the life cycles as a whole.

1. Introduction

The life cycle is one of the basic characteristics of an organism. By life cycle I mean the sexual reproductive cycle of an eukaryote, i.e., the alternation of meiosis and syngamy that leads to the alternation of a haploid and a diploid phase. A given cycle is characterized by the relative position of the two processes (meiosis and syngamy) that make the transition between the two phases (haploid and diploid), and by the duration and importance of the two phases. Many combinations are possible, ranging from one process immediately following the other, where one phase is reduced to a single cell, to the two processes being separated in time and space by many cell divisions (Fig. 1). When meiosis directly follows syngamy, the diploid phase is reduced to the zygote, and the organism grows mostly as a haploid. On the contrary,

---

1991 Mathematics Subject Classification. Primary 92D15, 92D40.

This work was supported by a grant from the Swiss National Funds to S.C. Stearns, and by a grant from the Roche Fondation to V.P.

The final version of this paper will be submitted for publication elsewhere.

when syngamy directly follows meiosis, the haploid phase is reduced to the gametes, and the organism grows mostly as a diploid. When meiosis and syngamy are far apart, both phases are individualized, one phase possibly growing out of the other. If we are to understand how the sexual life cycle evolved into the wide diversity we can still observe, we have to understand which cycle is favored under which conditions.

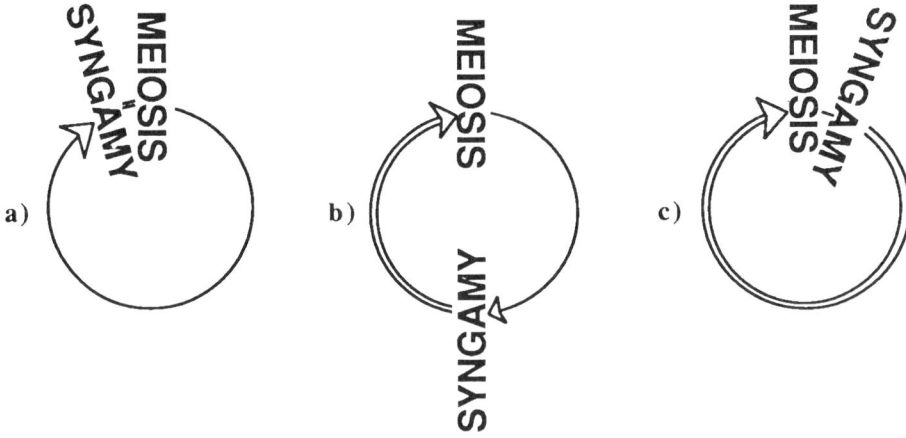

Figure 1.   The life cycle continuum. The cycles where meiosis immediately follows syngamy (a) or syngamy immediately follows meiosis (c) are the two extremes of a continuum represented by (b) where the two processes are separated in space and time along the cycle. Single and double lines represent haploid and diploid phase respectively.

When is it good to produce haploid offspring? When is it better to fuse and become diploid? How long should one stay as a diploid and undergo selection at that stage? Understanding the evolution of life cycles would answer all these questions. The life cycle of an organism describes the succession of stages and processes the organism goes through. Thus, the life cycle is a specially complex character. In animals, a single individual passes through the entire cycle. In many plants and algae, on the other hand, different individuals represent different parts of the cycle. The different parts that compose the life cycle are tightly linked and directly connected; they all play a role. The life cycle of an organism is an important trait, and as such we would like to understand how it gave the different cycles we can see in nature at present. To understand the evolution of life cycles, two kinds of empirical evidence are available: comparative and experimental data.

Comparative data The diversity of cycles we observe now is the result of evolution since the appearance of the eukaryotic sexual cycle. Broad evolutionary trends are visible, from which we can get insights into the evolutionary pressures acting on the life cycle. The comparative evidence will not be discussed in the present paper; see Raper & Flexer (1970) and Bell (present volume).

Experimental data Here we try to understand the properties of each component of the cycle, before putting the components back together to understand their interactions, that is, the cycle as a whole. The rest of this paper will be devoted to discussing the experimental approach.

2. What could be the difference between
a haploid and a diploid?

Different life cycles differ in two respects: the timing of the processes (meiosis and syngamy); the length of the phase between them, that is, how much of the complete cycle is spent as a haploid or a diploid. To understand the relevance of the latter characteristic of the cycle, we need information about the properties of haploids and diploids. The comparison of haploids and diploids is the starting point of the experimental study of life cycles. It does not give all the necessary clues, but it is a prerequisite to the understanding of life cycles. I will first review the hypotheses that have been proposed to compare haploid and diploid individuals (for a complete review of the hypotheses concerning the evolution of life cycles, see Valero et al. 1992).

Being diploid rather than haploid has two kinds of consequences. The first concerns the information content of the genome. Because it results from the fusion of two haploid cells, a diploid cell has a genome twice as large as that of a haploid cell. This has important consequences for the genetics of haploids and diploids, both with respect to the alteration of the genetic information, and with respect to the expression of genes. The other kind of consequence has to do with the number of DNA molecules present in the cell nucleus. A haploid cell has half the DNA to replicate at each cell division. The amount of DNA is likely to alter the size of the nucleus and the cell, thereby modifying the ecological properties of the cells. These consequences can be interpreted in various ways, leading to different hypotheses about how they influence the evolution of life cycles. Of course, genetic and ecological aspects are two sides of the same coin. The enzymes that process nutrients are coded by the genes, and in that sense the ecological aspect depends on the genetics. On the other hand, the genetic information is borne by DNA molecules packed in cells, and is not independent from matter.

2.1. Genetic consequences and hypotheses.

2.1.1. Information alteration. A diploid has two versions of each gene. The two versions may be identical or different. If the two versions are identical, the amount of information is the same for haploids and diploids, but redundancy is increased in diploids. This redundancy seems to favor diploids in several different ways.

(a) Double-strand damage reduction. Double-strand damage can be repaired if a second DNA molecule is present. In contrast to mutation (such as base substitution), DNA damage prevents the gene from being expressed. Enzymes that repair the damaged DNA need another intact molecule as a template when both strands of the double helix are affected. The template molecule is available at the diploid stage, but not at the haploid stage. This makes the diploids less sensitive to mutagens that damage DNA. See Michod (present volume) for a complete account of this theory.

(b) Masking of deleterious alleles. When one copy of a gene is altered, that copy is likely to be unable to fulfil its function. In that case, the spare copy can take over, sheltering the diploid from the expression of deleterious mutations. This is the most commonly-used argument for an advantage to diploidy. The extent to which diploidy is advantageous depends on the amount of masking, i.e., how much the "bad" allele is hidden by the "good" one. The amount of masking is usually referred to as the degree of dominance of the deleterious allele. The range of the degree of dominance which favors diploidy depends on the level of selection (Perrot, Richerd & Valéro 1991). When comparing a haploid and a diploid, both bearing a deleterious allele at a given locus, it is obvious that the diploid suffers less from the mutation as long as the mutated allele is not completely dominant over the normal allele. When comparing haploid and diploid sexual populations in mutation-selection balance, the mean fitness reduction due to a single recurrent mutation is the same in both populations for a completely recessive mutation. However, if the mutation is expressed, even slightly, at the heterozygous state, i.e., is somewhat dominant, the fitness reduction is larger in the diploid than in the haploid population. In this case, haploidy is favored. Variations on this theme have been extensively studied by population geneticists (Crow & Kimura 1965, Kondrashov & Crow 1991, Perrot, Richerd & Valéro 1991, Bengtsson 1992, Otto & Goldstein 1992, Otto in this volume). Their models show the conditions under which diploidy is advantageous. Despite its obviousness, masking does not always favor diploidy.

(c) Innovation using the spare copy (Lewis & Wolpert 1979). Since diploids have a spare copy of each gene, modifying one copy does not necessarily jeopardize the survival of the bearer. In that sense, diploids have more freedom to try out new gene functions. If one of the novelties happens to be useful, the heterozygote bearing both the old allele fulfilling the old function and the new allele performing a new task is at an advantage. The heterozygote advantage (overdominance) can then be fixed by duplicating the new allele, thus keeping the two functions permanently in the genome by preventing segregation. To be able to innovate in the same way, a haploid would have to duplicate a gene first, and then try out new versions. Lewis and Wolpert (1979) formalized this argument in a model that compares isolated haploid and diploid populations.

2.1.2. Mutation rate. To have twice as many molecules to replicate is a double-edged sword: diploids can potentially make twice as many mistakes per genome, and therefore have twice as high a mutation rate with respect to mutations caused by inaccuracy in DNA replication. On the one hand, it potentially generates twice as many advantageous mutations. On the other hand, it potentially generates twice as many deleterious mutations. Advantageous and deleterious mutations are thought to have different degrees of dominance. Data for various organisms (*Drosophila* (Simmons & Crow 1977), *Chlamydomonas* (Orr 1991)) show that deleterious mutations are generally more or less recessive, although usually not entirely so. The theoretical explanation concerns the kinetics of enzymatic reactions (Kacser & Burns 1981). The most efficient protein, that is, the product of the normal allele, will suffice to fulfil most of the gene function in the metabolic pathway, thus masking the failure of the mutated enzyme. This explains why deleterious mutations are generally recessive, and why a mutation improving gene function is expected to be more or less dominant. If advantageous mutations are generally dominant, this could give an additional advantage to diploidy. We have seen that the masking of deleterious alleles does not necessarily give an advantage to diploidy. Whether advantageous mutations actually favor diploidy is still open to theoretical investigation.

2.1.3. Capacity for complexity. When comparing organisms belonging to the same group (e.g., brown algae), one notices that the highest level of structural complexity is achieved in the diploid phase. It has been suggested that diploidy could be necessary for extensive tissue differentiation, although clear mechanisms are unknown (Raper & Flexer 1970, Maynard Smith 1978). As proposed by Raper & Flexer (1970), "diploidy may confer the *phylogenetic* capacity to evolve differentiated tissues" (their italics). This does not rule out the possibility of complex haploid organisms, like the haploid males of the hymenopterans, but could explain why they appeared only after a long evolution as fully diploid animals.

2.2. Ecological consequences and hypotheses.

2.2.1. C-value consequences. The DNA content of a cell (C-value) is positively correlated with cell size and with cell cycle length across eukaryotes (Cavalier Smith 1978). Following this line of argument, Cavalier Smith suggests that haploidy is favored when small size is advantageous, and that diploidy is favored when large size is advantageous. Cavalier Smith views the amount of nuclear DNA as the primary factor controlling cell size. His view is supported by a fair amount of data reviewed in his paper (Cavalier Smith 1978). The main objection to this view as an important factor in the evolution of life cycles is raised by Lewis (1985, see below).

2.2.2. Nutrient-saving hypothesis. Lewis points out that DNA is costly. In environments where nutrients such as phosphorus and nitrogen are scarce, DNA is an expensive way to set the cell size. Reducing DNA content is an efficient way to reduce the costs of cell division. Haploids have half the DNA to replicate, and thus need fewer nutrients per cell division. When the environment is poor in nutrients, haploid cells should be able to divide when diploid cells cannot (Lewis 1985).

2.2.3. Within the same cycle, haploids and diploids occupy different ecological niches. The hypotheses reviewed so far favor either one or the other phase. With this line of thought, one expects to observe life cycles in which either the haploid or the diploid phase is extended and the other phase reduced. Life cycles in which both phases are of comparable size require a different explanation. Because the two phases have different properties, each could correspond to a different set of selection pressures, that is, to different niches. The previous ecological hypotheses refer to comparison among taxa at any level of organization. It is easy to adapt them to comparison of haploids and diploids belonging to the same species. The niche hypothesis does not generate predictions about the advantages of haploidy or diploidy. It simply points out that there should be some selective differences between haploid and diploid individuals of the same life cycle for such a cycle to be maintained. It is difficult to imagine that a cycle where there is an alternation of haploid and diploid generations is in a neutral equilibrium. There must be selection pressures that maintain both phases. This hypothesis is usually proposed by botanists (Stebbins & Hill 1980, Keddy 1981), and has been recently investigated by Jenkins (1993).

3. Experimental comparisons of haploids and diploids: how do they actually behave?

The purpose of the experiments described below is to compare the behavior of haploids and diploids under various environmental conditions. The goal is to identify differences in behavior caused by the difference in ploidy level. Nevertheless, if we are to look at ecological differences, we should try to control for genetic differences. If we look at genetic differences, we should try to control for ecological differences. To limit the genetic differences, we would like to work with haploids and diploids belonging to the same organism, which requires an organism that exists both in the haploid and in the diploid state. The primary difficulty is to find an appropriate organism. Most of the organisms we are familiar with retain only the diploid phase in their cycle (animals, higher plants). Two possibilities have been exploited. Asexual organisms such as *Aspergillus niger* exists in two "versions," haploid and diploid, without proper syngamy or meiosis (Fig. 2). Alternatively, organisms like yeast and algae, that have a haplo-diploid life cycle, have been used to compare the two ploidy levels which alternate within the same life cycle (Fig. 1b). Some comparisons between haploids and diploids involve very artificial systems (e.g., animal cell cultures, plant protoplasts), and were not originally designed to understand the evolution of life cycles. They give insights into the functioning of haploids and diploids, but because they are concerned with very unusual parts of the reaction norm of the organism, they may have little relevance to the evolution of the life cycle of that organism. The available evidence is summarized in Table 1.

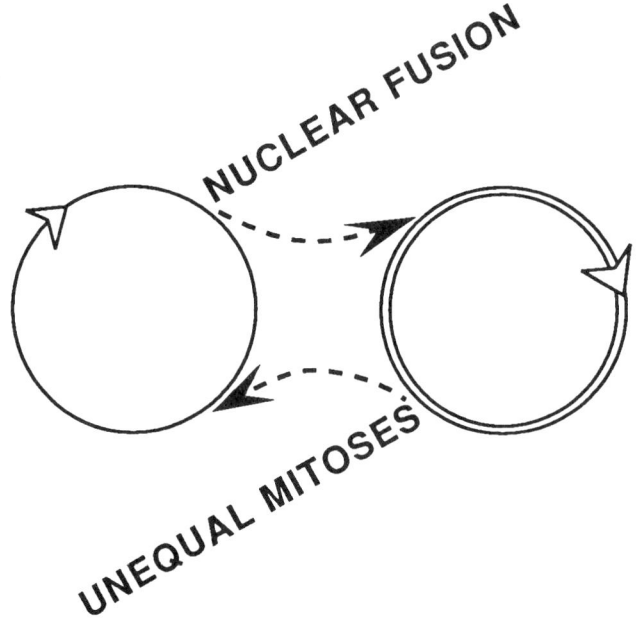

Figure 2.   The life cycle of *Aspergillus niger*. Both haploid (single line) and diploid (double line) states are stable. The transition between the two states occurs by fusion of two nuclei in the hyphae (from haploid to diploid state), and by loss of chromosomes by unequal mitoses (from diploid to haploid state)

3.1. Experiments dealing with the genetic consequences of being haploid vs. diploid.

3.1.1. Degradation of the genetic information: DNA damage and deleterious mutations. Diploids are expected to suffer less damage than haploids when exposed to a mutagen. This might be due to DNA repair, as diploids are expected to repair damage better then haploids, or to the masking of deleterious mutations, because if one allele is unable to carry out its function because of damage, the other allele is likely to be functional and can take over. The increased resistance of diploids to mutagens has been verified by comparing the reaction of haploids and diploids to exposure to a mutagen in yeast and in the red alga *Gracilaria verrucosa*.

In the budding yeast *Saccharomyces cerevisiae*, many experiments have been done to compare the action of DNA-damaging agents such as X-rays. In the older studies (e.g., Laskowski 1960, Mortimer 1958), the authors show that diploids survive irradiation better than haploids. In later work (e.g., Luchnik, Glaser & Shestakov 1977) the authors show that diploids are able to repair double-strand breaks in their DNA.

G. verrucosa is a red alga whose cycle involves an alternation of haploid and diploid identical individuals. In the laboratory, juveniles of both phases are exposed to heavy doses of UV light (Destombe & al. 1993). After 3 weeks, diploids grow better than haploids. What is measured here is the ability of irradiated cells to divide. The comparison involves two sets of individuals, not two populations evolving towards mutation-selection balance. In these circumstances, the masking of deleterious alleles, as well as DNA repair, favors diploids. It is difficult to distinguish between the two interpretations.

No experiments so far have been designed to test the population genetics models of the evolution of diploidy. There are many problems that hinder such a test. Firstly, the primary assumption made by all the models is that haploids and diploids would have the same fitness in the absence of deleterious mutations. This is of course a reasonable hypothesis, because the models deal with the genetic aspect of the process. Nevertheless, real organisms also have ecological constraints that make it difficult to isolate genetic factors alone. Secondly, the models describe sexual populations in equilibrium. This condition is difficult to fulfil in experimental populations, mainly because it is difficult -not to say impossible- to have sexual populations of the same organism following different cycles. So far, we have not managed to manipulate the cycle itself.

3.1.2. Improvement of the genetic information: advantageous mutations. Paquin & Adams (1983) compare the number of advantageous mutations accumulated by haploid and diploid strains of the budding yeast *S. cerevisiae*, maintained separately in chemostat culture with purely asexual reproduction. In this system, an advantageous mutation is detected by the appearance of a new clone that invades the chemostat (a phenomenon known to microbiologists as periodic selection). Strains are isolated at intervals from the culture in the chemostat, and are tested against strains previously isolated from the chemostat. Because the strains are asexual, any mutation which gives an advantage to the diploid clone in which it appears must have an effect at the heterozygous state. Their results show that the average fitness increment due to one mutation is the same in haploid and diploid strains, suggesting that the mutations are fully dominant in the diploids. Moreover, the diploid strains accumulate on average 1.6 times more advantageous mutations than the haploid strains per cell generation. This does not adequately answer the question about the relative mutation rate of haploids relative to diploids. The difference in mutation rate which is observed can be interpreted as a measure of the percentage of advantageous mutations that are more or less dominant. Assuming that the diploid mutation rate is double that of the haploids, about 80% of the mutations must be more or less dominant. There are questions left unanswered with this experiment. A mutation rate potentially twice as large would generate twice as many deleterious mutations in the diploid strains. Would that cost be counterbalanced by the advantageous mutations? What would be

Table 1: Hypotheses and experimental tests. Reported here are the working hypotheses and their tests. In the outcome column, H stands for haploids and D for diploids. The > or = signs indicate in what phase the measured character was larger.

Question	Organism used (ref. number)	Experimental conditions	Character measured	Outcome	Answer
does diploidy allow double strand damage repair? and masking of deleterious alleles?	G. verrucosa (1) S. cerevisiae (2)	haploid and diploid cells culture exposed to mutagen; grown in isolation	cell division rate	D>H	yes
is the adaptive mutation rate larger in diploids? (Paquin & Adams 1983)	S. cerevisiae (3)	haploid and diploid strains grown in chemostat; in isolation	advantageous mutation rate	D>H	yes
do haploids grow better in poor environment? (Lewis 1985)	G. verrucosa (1)	haploid and diploid juveniles grown in sea water or enriched sea water	individual growth	H > D in poor environment D > H in rich environment	yes
Are diploids "double haploids"	S. cerevisiae (4)	haploid and diploid strains grown separately in chemostat in different environments (minimum media)	growth rate	H = D	yes
		same strains as before, competing in the same chemostat, in same environments as before	ability of a strain to invade	H = D in 2 environments H > D in 1 environment	yes
Are diploids "double haploids"	S. cerevisiae (5)	haploid and diploid strains grown separately in chemostat on rich complete medium	growth rate	D>H	no
		mixed culture of haploid and diploid strains	ability of a strain to invade	D>H	no
Does a diploid strain outcompete its haploid parents?	A. niger (6)	haploid and diploid strains grown together on minimum medium	increase in frequency	H > D	no

References: (1) Destombe & al 1993; (2) Laskowski 1960, Mortimer 1958; (3) Paquin & Adams 1983; (4) Adams & Hansche 1974; (5) common yeast people observation; (6) Perrot, unpublished data the result of competition between the haploid and diploid strains that have accumulated advantageous mutations?

The hypothesis proposed by Lewis and Wolpert (1979), that spare copies of the diploids can experiment with potentially advantageous new mutations, is difficult to test experimentally, for the same reasons as the other population genetics models. Here, indirect evidence from molecular biology data could help. The number of gene families that have evolved by gene duplication is likely to be greater in organisms with a developed diploid phase than in organisms with a reduced diploid phase.

3.2. Experiments dealing with the ecological consequences of being haploid vs. diploid. To test these hypotheses, it is necessary to use individuals as genetically identical as possible, and to measure differences in growth. This can be done either by isolating haploids and diploids in the same conditions, or by growing haploids and diploids together. The former technique yields an estimate of the intrinsic properties of the two stages, whereas the latter yields an estimate of the competitive abilities of the two stages.

3.2.1. Nutrient scarcity vs. cell size (Lewis 1985, Cavalier Smith 1978). The hypotheses proposed by these two authors lead to the following predictions. Haploids are expected to grow faster (or divide more often) than diploids. According to Cavalier Smith (1978), this should hold regardless of the environment. According to Lewis (1985), this should hold in a poor environment, diploids growing as well as haploids (or possibly better than haploids) in a rich environment. These two hypotheses have been tested by measuring the growth of haploid and diploid juveniles of the red alga *G. verrucosa* (Destombe et al. 1993). Juveniles are grown in natural sea water and in enriched sea-water medium. The growth is measured after 3 weeks of cultivation. The results show that haploids grow better in the poor environment, and that diploids grow better in the rich environment. These results support the nutrient scarcity hypothesis of Lewis (1985).

3.2.2. Cell size and ploidy (Cavalier Smith 1978). The correlation between cell size and cell DNA content is well-established across eukaryotes. Nevertheless, this relationship is not universal at a smaller scale. Adams & Hansche (1974) measure the size of haploid and diploid yeast (*S. cerevisiae*) cells grown in 4 environments. In 3 environments, diploid cells are, as expected, larger than haploid cells. In one environment, however, the two types of cell are exactly the same size. The cell size ratio is not correlated with the competitive abilities of the strains (see below). This result indicates that unicells such as yeast are able to modify their size substantially in different environments without necessarily changing their competitive abilities. This is inconsistent with Cavalier Smith's view (1978) that the evolution of haploidy and diploidy is driven by the size of the organisms, at least at a microevolutionary scale.

3.2.3. Niche hypothesis. When both phases are individualized, the haploid and the diploid phase are expected to occupy different niches. When the two phases have very different morphologies, one expects the morphology of each phase to be adaptive. The alternation of blade (haploid phase) and crust (diploid phase) morphs in algae is a well-studied example of this phenomenon. Each morph is adapted to a different set of environmental conditions that alternate during the year. The crust successfully resists heavy grazing pressure, and the blade efficiently deals with strong competition (Littler & Littler 1980). The differences in the niches, if any, are more subtle when the morphological differences between phases are less clear. The phases can be very similar to the extent that one needs to examine the chromosomes and the sex organs to distinguish them. In these cases, people have looked for differences between haploid and diploid individuals that could explain the maintenance of the two phases in the cycle. This has been done with some species of red algae (*Polycavernosa debilis, Iridea*), where various characters have been measured (grazing pressure, primary productivity, dry weight calorific values, organic content). None of these characters differed between haploids and diploids (Hannach & Santelices 1985, Littler, Littler & Taylor 1987).

3.2.4. Competition experiments: "l'enfer, c'est les autres" (Sartre, 1942). Haploids and diploids can be compared when they interact with each other by competing for the resources present in the environment. The amount of resources available may depend on the activity of other organisms in the same environment. The amount of resources present in a grape is limited; what is available for a yeast is what has not yet been used by the other organisms growing there. The outcome of competitive interactions may affect the evolution of life cycles. This argument probably does not apply to the multicellular red algae. The marine environment in which they live might well be poor in nutrients, but nutrient limitation is unlikely to be influenced by neighboring algae. The outcome of competition between haploids and diploids has been studied using two microbial systems, the bread mold *Aspergillus niger* and the budding yeast *S. cerevisiae*.

(a) <u>Competition between haploid and diploid strains of *Aspergillus niger*</u> (Perrot, unpublished data). *A. niger* is an ascomycete fungus that forms a mycelium. Its development begins with the growth of a mycelium from a germinating spore. The mycelium later produces a large number of air-borne spores. The standard sexual cycle of the Ascomycetes is unknown in *A. niger*. A special kind of genetic exchange occurs, known as parasex. The transition from haploidy to diploidy happens by fusion of two haploid nuclei in the mycelium. The return to haploidy is made by a series of unequal mitoses. The resulting aneuploids seem very unstable and are not observed: only haploids and diploids are stable, and both produce spores. The formation of a diploid is a rare event. To isolate diploids, spores are screened by a complementation test between two haploid strains, each having a different growth requirement (i.e., bearing a different auxotrophic marker). The only spores that can grow on minimal medium are the diploid spores resulting from the fusion of the two haploid strains. The diploid strain is therefore heterozygous for at least two loci. One of the haploid strains I used also bore a spore-color marker. The marked haploid strains were derived from the same wild-type lab strain, with the markers introduced by UV mutagenesis. The strains are therefore likely to differ for other genes for which they have not been selected. A diploid strain was put in competition against both of its parental haploid strains by inoculating the medium with a suspension of spores with the competing strains in equal proportions. When the cultures sporulated, the production of viable spores by each strain was estimated. The results showed that the diploid strain was always out-competed by its haploid parents.

From the standard genetic theory on the evolution of diploidy, diploids should be at an advantage because they mask deleterious mutations. UV mutagenesis is likely to have generated different deleterious mutations in the irradiated strains, so than when two such strains are combined in a diploid, the diploid gains a genetic advantage over its parents. The competition result I obtained showed that there was also an ecological cost to being diploid, and that it was larger that the genetic gain in our strains. *A. niger* is normally more often in the haploid than in the diploid state. 200 strains isolated from Indonesia were all haploid, although 3 out of 10 strains isolated from the Netherlands were diploid (Hoekstra & colleagues, pers. comm.).   A possible explanation for my result is that *A. niger* has evolved as a haploid organism, and for this reason is more efficient as a haploid than as a diploid.   If the past evolutionary history of an organism affects the relative success of haploids and diploids, then the ability to change from haploid to diploid dominance in the cycle will be affected by evolved constraints as well as by short-term selection.   Another explanation is that haploids out-competed the diploid under the experimental conditions I used, but the outcome of competition may be different in different conditions.   For example, diploids might be favored in a medium richer than the one I used.

(b) <u>Are diploids "double haploids"?</u>   The hypothesis proposed and tested by Adams & Hansche (1974), using the budding yeast *S. cerevisiae*, is a refinement of Lewis's (1985).   They expect the mechanism of nutrient limitation to influence the outcome of competition between haploids and diploids.   When growth is limited by the utilization of nutrients, then haploids and diploids should grow at the same rate.   If nutrient intake rates are not limiting, haploids and diploids deplete the medium at the same rate.   If a diploid is simply a double haploid, it has twice as much DNA to replicate but also twice as many enzymes to do so, and its utilization efficiency will be the same as that of a haploid.   On the other hand, when growth is limited by the rate of nutrient uptake, haploids are expected to do better than diploids, because haploids require less nutrient per cell.   This hypothesis contradicts the common wisdom of yeast geneticists, that mixtures of haploids and diploids always become dominated by diploids.   Haploids are also known to grow more slowly than diploids.

Yeast is grown in chemostat, and two different environmental conditions are used, in which two different factors can affect cell growth.   In the first environment, nutrient use is supposed to be the factor limiting growth.   In the second environment, nutrient intake is supposed to limit growth.   The authors measure first the maximum reproductive rate of the strains in the two media when all necessary nutrients are present in excess.   There is no significant difference between haploids and diploids in either medium.   The authors then measure the ability of a strain to increase in frequency in a culture where growth is limited when introduced at a frequency of 10% of the resident strain.   In the first medium, the diploid strain is not able to increase significantly in frequency when introduced into a chemostat containing the haploid strain.   In the second environment, the haploid strain is able to increase in frequency when introduced into the diploid culture.   These results support the hypotheses proposed by the authors.   It is unfortunate that the reciprocal experiment was not done (or reported), i.e., introducing haploids into the diploid culture in the first environment, and introducing diploids into the haploid culture in the second environment.   This would have ruled out the possibility of frequency-dependent selection affecting the results.

In these experiments, diploids are at a disadvantage compared to haploids, at least under some environmental conditions.   Under other conditions, haploids and diploids seem to grow equally well.   *S. cerevisiae* is a "diploid" yeast, in the sense that it becomes diploid if given the possibility, the haploid stage being only the resting stage (spores), and, in heterothallic strains, the waiting stage until a member of the opposite mating type is found.   The observation that diploids are not better than haploids inconsistent with the argument that the most efficient phase is the one more exposed to selection, where more evolution has taken place.

It is possible that the conditions in which haploids do better than diploids are not normally encountered by yeast.   The usual observation that diploids out-compete haploids in mixed cultures is likely to have been made when using a rich medium, although this has not yet been tested formally.

4. Conclusion

Haploids and diploids may differ in two ways: in their genetics, and in their ecology. The theory of the genetic aspects is well-developed, and is fairly general. Experimental data concerning the genetic differences are needed in order to assess the importance of the genetic aspects in the evolution of life cycles. The ecological hypotheses, by contrast, involve much more restrictive assumptions. More data are necessary to improve the theory and refine its predictions. The interaction with the environment may play a crucial role in the ecology of haploids and diploids, each environment creating a different set of conditions that substantially modify the outcome of the theoretical analysis.

As shown by the data presented above, we have begun to get insights into the properties of haploids and diploids. Both the genetic and the ecological aspects of the question have been addressed. It seems well-established that diploidy is advantageous in highly mutagenic environments. This advantage may be less clear under normal conditions. The advantages to diploidy caused by the masking of deleterious mutations as well as by the enhanced ability to acquire advantageous mutations have not yet been experimentally demonstrated. There is a body of data suggesting that haploids grow better in poor environments whereas diploids grow better in rich environments. That haploids are favored in poor environments is expected by Lewis (1985). That diploids are favored in rich environments indicates that diploids are better than just "double haploids" as expected by Adams and Hansche (1974). When resources are abundant, diploids are more efficient at taking advantage of them than are haploids. There is need for more experiments designed to confirm this speculation.

We have until now compared only the properties of haploids and diploids, not the properties of different cycles. In the experiments I have described here, haploids and diploids have been carefully kept in genetic isolation. During the course of the experiments, no sexual reproduction was allowed. It is very difficult to manipulate the cycle of an organism so that we can compare populations of the same organism following different cycles. One of the very few organisms with a variable life cycle is *G. verrucosa* (Destombe & al. 1989). Unfortunately, this red alga grows slowly and is difficult to culture in the lab, so until now no artificial population experiments have been set up with this organism. Experiments with *S. cerevisiae* that constrain different sexual populations growing in parallel to a haploid or a diploid phase are very time-consuming because it is difficult to maintain haploidy in a sexual population of this yeast. Because of the general lack of organisms polymorphic, or even variable, for life cycle and our inability to conveniently manipulate the life cycle itself, we must be satisfied with comparative data. Experiments to date have used only asexual populations. This means that life cycles *per se* have never truly been compared, but rather that different parts of the same life cycle have been compared (see Fig. 1). Comparative data should allow clarification of the ecological conditions that favor the different cycles. It is unfortunately not clear how flexible life cycles are. The very fact that organisms polymorphic for the life cycle seldom exist certainly suggests that a given cycle is most appropriate for the conditions experienced by a single organism or by its ancestors. Perhaps we, as animals, are diploid simply through phylogenetic constraints. Our progenitors experimented with diploidy in the distant past, and have taken a road of no return.

REFERENCES

Adams, J. & Hansche, Population studies in micro-organisms. I Evolution of diploidy in *Saccharomyces cerevisae*.. Genetics **76** (1974), 327-338.

Bengtsson, B.O. Deleterious mutations and the origin of the meiotic ploidy cycle. Genetics **131** (1992), 741-744.

Cavalier Smith, T. Nuclear volume control by nucleoskeletal DNA, selection for cell volume and cell growth rate, and the solution of the C-value paradox. J. Cell Sci. **34** (1978), 247-278.

Crow, J.F. & Kimura, M. Evolution in sexual and asexual populations. Am. Nat. **99** (1965), 439-449.

Destombe, C., Godin, J., Nocher, M., Richerd, S. & Valero, M. Differences in response between haploid and diploid isomorphic phases of *Gracilaria verrucosa* (Rhodophyta: Gigartinales) exposed to artificial environmental conditions. Hydrobiol. (1993) in press.

Destombe, C., Valero, M., Vernet, P. & Couvet, D. What controls haploid-diploid ratio in the red alga, *Gracilaria verrucosa*. J. evol. Biol. **2** (1989), 317-338.

Hannach, G. & Santelices, B. Ecological differences between the isomorphic reproductive phases of two species of *Iridea* (Rhodophyta, Gigatinales). Mar. Ecol. Prog. Ser. **22** (1985), 291-303.

Jenkins, C.D. Selection and the evolution of genetic life cycles. Genetics **133** (1993), 401-410.

Kacser, H. & Burns, J.A. The molecular basis of dominance. Genetics **97** (1981), 639-666.

Keddy, P.A. Why gametophytes and sporophytes are different: form and function in a terrestrial environment. Am.. Nat. **118** (1981), 452-454.

Kondrashov, A.S. & Crow, J.F. Haploidy or diploidy, which is better? Nature **351** (1991), 314-315.

Laskowski, W. Inaktivierungsversuche mit homozygoten Hefestämmen verschiedenen Ploidiegrades. Z. Naturforschg. Teil B **15b** (1960), 495-506.

Littler, M.M. & Littler, D.S. The evolution of thallus form and survival strategies in benthic marine macroalgae: field and laboratory tests of a functional form model. Am. Nat. **116** (1980), 25-44.

Littler, M.M., Littler, D.S. & Taylor, P.R. Functional similarity among isomorphic life-history phases of *Polycavernosa debilis*. J. Phycol. **23** (1987), 501-505.

Lewis & Wolpert. Diploidy, evolution and sex. J. theor. Biol. **78** (1979), 425-438.

Lewis, W.M. Jr. Nutrient scarcity as an evolutionary cause of haploidy. Am. Nat. **125** (1985) 692-701.

Luchnik, A.N., Glaser, V.M. & Shestakov, S.V. Repair of DNA double-strand breaks requires two homologous DNA duplexes. Mol. Biol. Rep. **3** (1977), 437-442.

Maynard Smith, J. The evolution of sex. Cambridge University Press, Cambridge 1978.

Mortimer, R.K.   Radiobiological and genetic studies on a polyploid series (haploid to hexaploid) of *Saccharomyces cerevisiae*. Rad. Res. **9** (1958), 312-326.

Orr, H.A. A test of Fisher's theory of dominance. Proc. Natl. Acad. Sci. USA **88** (1991), 11413-11415.

Otto, S.R. & Goldstein, D.   Recombination and the evolution of diploidy. Genetics **131** (1992), 745-751.

Paquin, C. & Adams, J. Frequency of fixation of adaptive mutations is higher in evolving diploid than haploid yeast populations. Nature **302** (1983) 495-500.

Perrot, V., Richerd, S. & Valéro, M.   Transition from haploidy to diploidy. Nature **351** (1991), 315-317.

Raper, J.R. & Flexer, A.S. The road to diploidy with emphasis on a detour. Symp. Soc. Gen. Microbiol. **20** (1970) 401-432.

Sartre, J.-P. Huis-clos. Gallimard, Paris 1942.

Simmons, M.J. & Crow, J.F. Mutations affecting fitness in *Drosophila* populations. Ann. Rev. Genet. **11** (1977), 49-78.

Stebbins, G.L. & Hill, J.C. Did multicellular plants invade the land? Am. Nat. **115** (1980) 342-353.

Valero, M., Richerd, S., Perrot, V. & Destombe, C. Evolution of alternation of haploid and diploid phases in life cycles. Trends Ecol. Evol. **7** (1992), 25-29.

ACKNOWLEDGEMENTS

Thanks to Jacqui Shykoff and Erika Bucheli for providing a shelter and a stimulating environment at the Mountain Lake Biological Station (Virginia) during the final phase of the writing of this paper. Thanks to Rustom Antia, Erika Bucheli, Lorenz Gygax, Sophie Richerd and Jacqui Shykoff for constructive comments on the manuscript, and to Graham Bell for correcting the English.

ZOOLOGISCHES INSTITUT DER UNIVERSITÄT BASEL, RHEINSPRUNG 9, CH-4051 BASEL, SWITZERLAND.

E-mail address: perrot@urz.unibas.ch